The $100 Hamburger

Other McGraw-Hill Aviation Titles

The Illustrated Buyer's Guide to Used Airplanes, Fourth Edition
Bill Clarke

Airplane Ownership
Ronald J. Wanttaja

Aviation Computing Systems
Mal Gormley

The Aviation Fact Book
Daryl E. Murphy

Beyond the Checkride:
What Your Flight Instructor Never Taught You
Howard J. Fried

Flight Instructor's Pocket Companion
John F. Welch

Airplane Maintenance & Repair:
A Manual for Owners, Builders, Technicians and Pilots
Douglas S. Carmody

200 Best Aviation Web Sites
...and 100 More Worth Bookmarking
John Allen Merry

The $100 Hamburger

A Guide to Pilots' Favorite Fly-In Restaurants

John Purner

McGraw-Hill

New York San Francisco Washington, D.C. Auckland Bogotá
Caracas Lisbon London Madrid Mexico City Milan
Montreal New Delhi San Juan Singapore
Sydney Tokyo Toronto

Library of Congress Cataloging-in-Publication Data

Purner, John.
The $100 hamburger : a guide to pilots' favorite fly-in restaurants / John Purner.
p. cm.
One hundred dollars hamburgers.
ISBN 0-07-083714-7 (acid-free paper : pbk.)
1. Restaurants—Guidebooks. 2. Air-pilot guides. I. Title
TX907.P87 1998
647.95—dc21 98-16271
CIP

1 2 3 4 5 6 7 8 9 0 FGR/FGR 9 0 3 2 1 0 9 8

ISBN 0-07-083714-7

The sponsoring editor for this book was Shelley Carr, the editing supervisor was Jane Palmieri, and the production supervisor was Sherri Souffrance. It was set in Houston by Ronald Cecil.

Printed and bound by Fairfield/Quebecor.

This book is printed on recycled, acid-free paper containing a minimum of 50% recycled, de-inked fiber.

Contents

Acknowledgments

This book is made possible by the contributions of over 1,000 pilots worldwide. Each generously e-mailed notes to me about their favorite fly-in eating spots.

Introduction

Wilbur and Orville Wright made four successful flights on December 17, 1903. By the end of the day, the world had substantially and irrevocably changed.

The first privately owned light airplane belonged to the Wrights. Orville had given little consideration to the question of its utility. Wilbur the more impetuous of the two blurted out the obvious answer, "Orv, what say we fire this puppy up and fly to a great little restaurant I've heard about in Ocracoke? It has terrific burgers for only a nickel!"

"Will, are you daft? Have you considered the expense of operating our invention? It'll cost us $100 just to get to Ocracoke! So it's not a nickel hamburger. It's a $100 hamburger," scolded Orville.

"I know Orv, but it will be a real kick. Where's your spirit of adventure?" asked Wilbur.

It is possible that I am misinformed and this conversation between the Wrights never ***REALLY*** happened. I think it did!

One thing is certain, thousands of contemporary pilots are making burger runs every weekend. I am one of them. The biggest problem for all of us is finding fly-in restaurants.

The $100 Hamburger addresses this difficulty. It is a hungry pilot's guide to restaurants at or very near general aviation airports worldwide. If your ship is a Cessna, Piper, Mooney, Beechcraft or any other machine that flies and you're passionate about flying and eating, WELCOME HOME! This book is for you!

The $100 Hamburger is part of a project that began in 1995. The Internet was forcing its way into our consciousness. Clearly, all of our lives were to be changed by it in big and little ways. This genie of the next millennium offered a new and better way to communicate and share information. I decided to create a website for pilots who simply wanted to know where they could fly and buy a hamburger, nothing grander than that!

From the beginning, I noticed the "hit" counter of ***The One Hundred Dollar Hamburger*** website begin to climb. It was an immediate success. By late 1997, this on-line guidebook was receiving over 3,000 hits a day! The best news was the active participation of the flying community. Over one thousand pilots e-mailed their thoughts on restaurants they had recently visited. I tagged these Pilot Reports, or PIREPs, if you please! Others would e-mail news of important changes at these previously reported-on fly-in locations. Those I labeled AMENDMENTs. 100LL pricing reports were added in mid-1997. The best deals on fuel for man and machine could now be found instantly at http://www.tpwi.com, ***The One Hundred Dollar Hamburger*** website. Today, it provides more than 3,500 first-hand PIREPs and AMENDMENTs on over 1,000 fly-in cafes. In excess of 100 new reviews come in each month.

However, the website has a serious flaw. You can't take it with you.

I recently flew from Wharton, Texas, to Oshkosh, Wisconsin, for the Big Show! The flight was a casual affair. We would stop when we needed fuel and arrive when we got there. Adventure, not schedule, was to be served. About noon, my flying buddy said, "John, I'm hungry. Where shall we eat?" Certainly a fair question for him to put to me, the father of ***The One Hundred Dollar Hamburger*** website. I should be expected to know the location of the nearest and best fly-in burger cafe anyplace on the planet. I couldn't remember where all of those thousand hamburger spots were at that moment and still can't today. The website is terrific, but I needed something I could carry in the cockpit. For a few hundred years, the world has called that something a book!

To help you decide were to fly, I devised a comparative rating system based on location, ambiance, service and food. You'll find one to five burgers next to each restaurant name. Five indicating the VERY best and one meaning you won't starve but don't expect much. The most important rating factors to me are location and ambiance. ***The $100 Hamburger*** is a guide to fly-in food. There are usually better restaurants downtown, but if they aren't pilot accessible they don't belong in ***The $100 Hamburger***. Any clean, well-kept cafe with fresh paper placemats can be appreciated. A restaurant with a deck overlooking the flightline or the nearby ocean is something special! We eat with our eyes as well as our mouths.

There are tradeoffs in life. A directory, like a newspaper, becomes dated the moment it is written. A portion of the information it provides will change before a reader can take advantage of it. The website is continually updated, but it isn't portable. Each is imperfect. Together, they comprise a powerful information resource, portability from the book and timeliness from the website. The sum of the two is something I call a cyberbook. I believe the future of publishing is headed in this direction.

I hope ***The $100 Hamburger*** (book) becomes your cherished cockpit companion. Between editions, please feel free to use the updates available on ***The One Hundred Dollar Hamburger*** (website) and consider becoming a contributor. Use the Restaurant PIREP form in the back of the book, or use the Net: http://www.tpwi.com.

John Purner

Part 1

The United States

Andalusia (Andalusia Muni. - 79J)

Chart House

The Chart House is about four miles from the airport. It serves a very good and reasonably priced lunch buffet. Don't worry about transportation, the FBO has a courtesy car!

Dauphine Island (Dauphine Island - 4R9)

Seafood Galley

The Seafood Galley on Dauphine Island is wonderful. Next to the phone at the FBO is a placard with their number. Call and they will pick you up at no charge. Have a good meal!

Dothan (Dothan Muni. - DHN)

Airport Oasis Restaurant

At 8,400 feet, the runways are long enough here to land two, maybe three times. The best thing about this cafe is that it's right on the field. Fill up the ship with fuel and your passengers with food and be on your way.

Eufaula (Weedon Field - EUF)

The Oasis

Located in east-central Alabama on the expansive Lake Eufaula, this airport has an excellent seafood restaurant, bar and poolroom on the field. It has a genuine "honky tonk" feel and very good food. Mind the hours, Tues. through Sat. 05:00 PM - 09:45 PM.

Fairhope (Fairhope Municipal Airport - 4R4)

The Grand Hotel

The Grand Hotel, located on Mobile Bay and owned by the Marriott chain, is an excellent place to go. On Sundays they have a champagne brunch that is out of this world. The cost is about $15.00 plus gratuities per person. Of course, you pilots shouldn't participate in the champagne, but your passengers can. The grounds are beautiful and you can walk along Mobile Bay after dinner. Land at (4R4) and the FBO will call a hotel courtesy car for you.

Fayette (Richard Arthur Field - M95)

LB's Bar-B-Q

One of the best food stops in Alabama is at Arthur Airport (M95) near Fayette. Look on the Atlanta Sectional 50 miles west northwest of Birmingham. Check out LB's Bar-B-Q (205- 932-7100) located next to the Kentucky Fried Chicken. No auto needed here. Both restaurants are located within 50 yards of the ramp. The FBO is not opened on weekends so bring enough gas to get home.

Huntsville (Huntsville Intl. - HSV)

Sheraton Hotel Restaurant

The Sheraton Hotel in the main terminal has a nice restaurant, with good food, reasonable prices, and a view of the runways. It's only a very short hike from the FBO, Signature.

Green Bottle Grill

The Green Bottle Grill (205-882-0459) is an excellent restaurant with a variety of special dishes. This is a place to go to dine. As long as you are here, visit the U.S. Space & Rocket Center, fun for all ages. Try flying the MMU (Manned Maneuvering Unit) or take the bus tour of Marshall Space Flight Center.

Jasper (Walker County-Beville Field - JFX)

Hoyett's Place

Hoyett's Place of Jasper, Alabama, has good burgers and great catfish. It's about two miles from the airport. There is usually a nice airport car available. The food is good and the staff attractive.

Madison (Decatur/Athens - DCU)

Greenbrier Bar-B-Que

Fly into Decatur/Athens airport (DCU) and have lunch at Greenbrier Bar-B-Que. It's located several miles off the field. Clay, Cass, Stan or Nicole at Decatur/Athens Aero will let you use one of several airport cars. The food is good and the staff at DAAS very accommodating. They are the local Comanche specialists for those of you who fly REAL airplanes.

Pell City (St. Claire County Airport - PLR)

JJ's Restaurant

JJ's Restaurant and Lounge is near the St. Claire County Airport. It has a pilot-oriented ambiance and is run by Jim Gray, who also operates the Airport FBO. PLR is a nice airport with a 5,000-foot runway. The crew car is usually available. It's a cool airport with P51, T-6, biplanes and other neat planes. You can even skydive on weekends. (205-338-9456)

Tuscaloosa (Tuscaloosa Muni. - TCL)

Dreamland Bar-B-Q

Slabs are only $14.00. My wife and I flew down last weekend. We have made this trip a couple of times lately. Yes, its a 5 +++.

The Northport Cafe

The Northport Cafe, located in downtown Northport, has some of the best "home cooking" around. Use the courtesy car to make the five-minute drive into town for some outstanding vegetables, fish, beef and just generally good food. Best of all, it is inexpensive. Lunch is less than $5.00.

Circle Hot Springs (Circle Hot Springs - CHP)

Circle Hot Springs Lodge

This lodge serves great prime rib and other specialties. Lodging is also available year-round.

It is located 95 nm northeast of Fairbanks on the 135 degree radial at 67.5 DME off the Fort Yukon VORTAC, GPS coordinate N65.29.13 W144.36.65, deep in the interior of Alaska. Access is by wheels only.

Katmai National Park (Brooks Lodge – Floatplanes)

Brooks Lodge

This summer-only lodge offers some terrific "homemade" food. Everything from the main course to dessert is fantastic. The area boasts the largest brown bear population in the world and is truly an exciting place to explore.

It is located in the heart of the Katmai National Park on the 087 degree radial at 32 DME off the King Salmon VORTAC. The GPS coordinates are N58.33.18 W155.46.30. Access is by seaplane only.

Manley Hot Springs (Manley Hot Springs - MLY)

Manley Hot Springs Lodge

Great "grease burgers" and rustic atmosphere. It boasts relaxing hot springs and year-round lodging.

Manley Hot Springs is 70 nm west of Fairbanks on the 079 degree radial at 40.4 DME off the Tanana VOR/DME. The GPS coordinates are N64.59.85 W150.38.65. Access by wheels and floats.

Northway (Northway – ORT)

The Airport Cafe

We have flown to Alaska twice in the last four years and plan another trip this summer. The restaurant at the FBO in Northway makes great pies! We haven't had any of the other menu items, but they look great! Of course, to make this trip from the states costs considerably more than $100, but it is definitely worth it!

Sitka (Sitka Muni. - SIT)

Terminal Cafe

Believe it or not, the best peanut butter pie in the world is right here at the Terminal Cafe in Sitka. For good greasy burgers and other fare, a short trip to town from the man-made aircraft carrier style runway is in order. As a note, there are always more things open when the cruise ships arrive. Things really crank into full swing then!

Skagway, Alaska (SGY)

Dea's Restaurant

Just 79 nautical miles southwest of Whitehorse, Yukon, Skagway is nestled amidst the coastal mountains of the Alaskan panhandle at the end of the Taiya Inlet. Famous as the trailhead of the fabled Chilkoot Pass, where goldseekers lugged their ton of supplies over the mountain into Bennett Lake.

What makes this a great destination for the $100 hamburger is:

1. Its breathtaking beauty.
2. The strip is quite literally in town. Your own two feet are all you'll need for transportation.
3. The historic cemetery where many graves date back to the gold rush.
4. The train ride to the summit of the White Pass, very scenic but a little pricey.
5. The many historic buildings and quaint shops along Main Street.

I recommend Dea's Restaurant for fresh Alaskan halibut fish and chips.

Red Onion Saloon

The Red Onion Saloon offers hearty food in a frontier setting.

You'll work up an appetite just getting to Skagway! Coming from Whitehorse, access to Skagway VFR will depend upon the condition of the pass, which can often be solidly blocked with clouds. Be sure you have a good idea what the weather is doing before going. Check the Skagway winds, they can be a bit excessive and gusty at times.

It is important to be on time for the customs officer because he/she will be driving from the border to meet with you. They also clear the train passengers and will not wait if there is a train arriving.

I recommend flying direct Whitehorse to Carcross (FA4), then following the highway over the pass. Alternatively, fly over Bennett Lake following the railway until it intersects the highway.

The prevailing winds coming off the water dictate a right-hand pattern. This takes you along a mountainside. Turn base when you see the waterfall on the river. Traffic can be heavy during days when there are cruise ships in the dock. Return flight plans can be filed with Juneau FSS by dialing 1-800 WXBRIEF just as you do in the lower 48.

Tok (Tok Junction - 6K8)

Fast Eddy's

Tok, Alaska, has a good breakfast/dinner cafe right next to the airport. It is called Fast Eddy's. Try it. You will like it.

Chandler (Chandler Muni. - CHD)

Hangar Cafe

GREAT breakfast and burgers for very reasonable prices. It is a very small, place so there can be a wait. Food service stops midafternoon. Some interesting aircraft are kept at Chandler, and there is an FBO that rents aerobatic aircraft.

AMENDMENT

The "Hangar Cafe" is a good restaurant worthy of at least a three hamburger rating! It is a great place to go for breakfast. I don't think they are open beyond about 3:00 PM. Try one of their slow roll breakfast burritos.

Flagstaff (FLG)

Bun Huggers

Not on the airport but just a couple of minutes by courtesy car away is Bun Huggers. It offers great burgers and brew on the cheap. FLG is a high mountain airport with beautiful mountain views. Use caution for density altitude and turbulence.

Glendale (Glendale Muni. - GEU)

Glendale Airport Cafe

My experiences with the restaurant at Glendale Municipal Airport have been mixed. Prices were average, food was average and the hours have been spotty. The last two times I have gone over there, they were just closing, even though the normal hours would have had them open for several hours longer.

AMENDMENT

This place has by far the best fried fish and clam chowder I have EVER had. The locals all drive there and the airport lobby is usually packed with people waiting for their all-you-can-eat fishfry on Friday nights. The clam chowder is thick and packed with clams. Easy to fly into, the same controller works both tower and ground and all you have to tell him is, "Taxi Restaurant"!

I rate this four burgers on a scale of one to five. You pick the appropriate proportion!

Kingman (Kingman Muni. - IGM)

Kingman Airport Cafe

This restaurant is your typical small town joint. I recommend going simply for the art and antiques hanging on the wall. These things bring you back to a time long gone by. Old WWII uniforms and photos dot the walls. The breakfast food is good and the prices are cheap. I highly recommend the pancakes. They are very, very fluffy.

Marble Canyon (Marble Canyon - L41)

The Trading Post

Marble Canyon is about 18 miles southwest of Page, AZ. The trading post is just across the street from the airstrip. The runway is only 27 feet by 4,500 feet so it can be a little tricky. The food is good but the scenery is the best in the Southwest. If you cross the Grand Canyon from the south via one of the corridors, the views are breathtaking. Once at the north rim, it is about 30 miles to the airstrip. The Vermilion Cliffs rise to the west. When they join the Colorado River gorge you're there, the most incredible views in the country. This is the type of scenery that movies are made of. Don't miss it.

Mesa (Falcon Field - FFZ)

Falcon's Roost

Falcon's Roost is on the airport and offers inexpensive food, very good hamburgers and daily specials. I had corned beef and cabbage on St. Patrick's Day. It was great and cost less than $5.00. The "Chas" burger is highly recommended. It was named after one of the well-known local instructors. There is also a bar that is a great place for a cheap "cold one" after a great day of flying. Falcon Field has some interesting homebuilt and classic aircraft hangared there, as well as an F-1 air racing team. The Champlin Fighter Museum is right on the airport.

Anzio Landing

There is also a very good Italian restaurant at FFZ called Anzio Landing. It is at the most eastern edge of the airport. Prices are moderate to expensive. The food is usually excellent. I think they are currently just opened for dinner. Also check out the wacky homebuilt "flying saucer" in front of the restaurant. A Mesa local built it and "flew" it, but he claims it scared him so he has loaned it to the restaurant as an attraction. It is based on the principal of a Frisbee!

AMENDMENTS

Anzio Landing is named after Anzio, Italy. It's Italian cuisine. Prices are moderate, food and service are very good, and it's an excellent dinner fly-in location. Reservations are recommended. Taxi up and park literally at the restaurant door. One side of the restaurant is all windows overlooking the approach end of RWY 22, a pilot's view for sure.

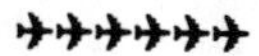

I would have to say the best fly-in restaurant is Anzio Landing at Falcon Field Airport in Mesa, Arizona. They have great daily lunch specials.

Mesa, AZ (William's Gateway Airport - IWA)

The Flight Deck Cafe 

William's Gateway Airport, located at the old William's AFB in Mesa, Arizona, has a great little hamburger joint called the Flight Deck Cafe. Since I live in Mesa, I drive there to eat. Almost every day at lunch there are several transient Air Force fighters parked out front while the jocks get a bite to eat.

Payson (Payson - E69)

The Crosswinds Cafe 

Very reasonably priced. My friends swear that they have the best biscuits and gravy in AZ. Once again, it's inexpensive. The country surrounding the airport is beautiful. It makes a very nice short "breakfast flight" from Phoenix.

AMENDMENT

Payson is one of the best places to go in Phoenix for breakfast! It is a short hop and is 10 to 15 degrees cooler, which can make a big difference in the summer. Great pancakes and lots of friendly people. They have a small air show there every April.

Phoenix (Deer Valley Muni. - DVT)

Deer Valley Food & Catering 

A favorite with all of my friends. Excellent breakfast and burgers. It definitely can be a little crowded on weekends, but is worth any wait. A very friendly airport.

AMENDMENT

One of our favorite fly-in restaurants is located in North Phoenix at the Deer Valley Airport. They offer great food at a reasonable price with a close view of the airplane action.

Prescott (Love Field - PRC)

Nancy's Skyway Restaurant

The restaurant on the field is definitely a greasy spoon but offers good burgers and the like at low prices. Seating is cramped. The place is generally very busy. There are airplanes everywhere you look, inside and out. Embry Riddle is located right next door. Arrive and depart the terminal building either through gate 1 or gate 2. Ya' gotta' see it!

Scottsdale (Scottsdale Muni. – SDL)

The Left Seat

The restaurant is in the main terminal building. The food is very good and is inexpensive to moderate.

AMENDMENT

If you get a burger at this cafe, be sure to get it with a slice of green chile! The chile is not hot, but it's very flavorful. Since I'm a native desert rat, I know my chilies!

Sedona (Sedona - SEZ)

Airport Cafe

My very favorite fly-in restaurant is at Sedona Airport. Oak Creek is without a doubt the most breathtaking area in AZ. The airport is on a small mesa and has a slight incline, so it can be a little intimidating at first for a low-time pilot. The restaurant has the best blueberry pancakes and is my personal pick for best biscuits and gravy. Prices are inexpensive to moderate. The traffic in and out of the airport can be very interesting as there are quite a number of celebrity types coming and going. Last time I was there, I saw several "bizjets" and the National Convention of Ercoupes—what an interesting mix.

Shagrue's

Shagrue's is in town, but the FBO will call a cab for you for the short ($5-10) ride. It is great for breakfast, lunch or dinner. Excellent food, lots of it, and moderate prices.

The Heartline Cafe

The FBO will call a cab for you for a short ($5-10) ride to the Heartline Cafe, which is owned and run by a top chef. Four star cuisine is offered at reasonable, for the quality, prices. There are only a handful of tables, the intimate setting demands reservations. Not to be missed!

Tucson (Avra Valley Airport - E14)

Airport Cafe

I have enjoyed breakfast numerous times at Avra Valley Airport north of Tucson. The help is friendly, the food good and the walk from the airport easy.

Tucson (Tucson International - TUS)

Tower Grill

I like the Tower Grill at the Intl. Airport. Excellent burgers and other stuff.

Tucson (Ryan Field – RYN)

Wings

There is a good fly-in restaurant at Ryan Field just west of Tucson, AZ. They serve lunch and dinner. The meals are reasonably priced. I consider the local folk eating there as a good sign of quality.

AMENDMENT

Wings has to be my first choice. What a great little place to fly in and eat. I wonder if they still feed the roadrunners by hand there?

Winslow (Winslow Muni. – INW)

Winslow Airport Deli

Can't remember the name of the place, but there's a small deli inside the FBO that makes a great BLT sandwich. Not fancy, but a great place to stop and fuel up.

AMENDMENT

The last time I checked (2-3 months ago), the restaurant inside the FBO building at the Winslow airport was closed. So give the airport a call before running over there for a burger! (602-289-2429)

The Famous Falcon Restaurant

The Famous Falcon Restaurant is a great place to eat when you fly into Winslow, AZ. It's on historic Route 66 and has been in business since the late 1940s. It was pretty nicely written up in the AZ Republic a few years ago. It's a family owned business and the food is homemade; really good chicken fried steak, made by hand. The restaurant has been visited by many celebrities. I spotted a picture of Kate Moss and Johnny Depp in the restaurant.

The restaurant is a way from the airport; I'd say over one mile. You have to just hoof it. It takes 15 minutes to walk to the restaurant. Ask any local who's standing on a corner in Winslow, AZ. They all know the way to the Falcon. If I remember correctly, you take the airport road to Highway 87, then hang a left and go under the old Santa Fe railroad tracks. The airport is south of the tracks and most of the town is north. The Falcon is at the northeast part of town.

Lots of old shops, the old train depot and hotel, you turn right at the first eastbound one-way street. There are large black billboards for the Falcon Restaurant that guide you there. You're less than halfway there by now. The Falcon will be on the left. The old Winslow Airport Cafe has been closed for a while so I sought to find another restaurant. This is a good one if you're into a little hike. Besides, it helps digest the food better with a little exercise.

Coming in from the West (Flagstaff – FLG), you pass near the meteor crater, quite a spectacular view.

El Torito Mexican Restaurant 

The El Torito Mexican Restaurant is very close to the airport. It serves some of the greatest food in the world. The people of Winslow and others across the nation enjoy their great cooking. It's north of the east end of the east-west runway. Take the airport road to Highway 87. Turn left and go north to the first left turn. Go west and it is the first restaurant on the right. It's a short distance by car but you can walk to the east end of the east-west runway and jump the fence and it is just across the street.

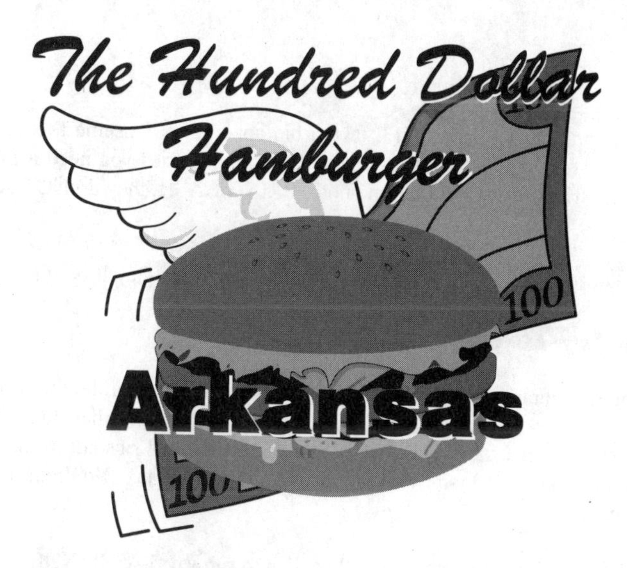

Fort Smith (Fort Smith Regional - FSM)

Jerry Neel's BBQ

One of my absolute favorite places to "lunch-and-back" is Jerry Neel's BBQ — lip-smackin' good BBQ pork, beef, or chicken with a generous menu of sides, including fried okra. I always get free cobbler (with ice cream, too!) after a meal. The FBO at Fort Smith, TAC Air, has a couple of courtesy cars for the short drive and will also give you a 30% off coupon to use at Jerry Neel's. What a deal! They are open every day but Sunday.

The Catfish Cove 

The Catfish Cove does an outstanding buffet on Sundays. They have more food than you would believe, including shrimp, BBQ beef, fried chicken, casseroles, fresh bread, homemade cobblers and catfish, of course!

Hot Springs (Memorial Field - HOT)

Hamilton House 

The Hot Springs National Park (HOT) airport staff will carry you to nearby restaurants. Discount Auto Rental at the airport rents cars for $10.00. Hot Springs is a beautiful place, truly first class. A $200 New York restaurant meal costs only $25.00 at the Hamilton House.

McClard's Bar-B-Que

The secret to McClard's Bar-B-Que is to get there early, no later than 11:30 AM, or wait until the noon rush is over. The bar-b-que is out of this world. In business since 1928 and only two miles from the airport, McClard's is not to be missed! It is rated as one of the best in the nation, so good that the President goes there when he's in town. It has served many other celebrities.

Try the tamale spread, but be warned, it'll take a big appetite to consume this tasty dish. The ribs and fries are a plate full. It's a slab of tasty ribs and so many fries you need a front-end loader to shovel them in. The fries are very good. One note, BRING CASH, they do NOT accept credit cards or checks.

Just ask the fellows around the airport to take you to McClard's; they'll be glad to do it.

Old Feed House

There is one more restaurant near the Hot Springs Airport. It is called the Old Feed House. This is an all-you-can eat freshwater fish and seafood restaurant with a few other items. Excellent catfish, seafood gumbo, chicken and dumplings, frog legs, and bar-b-qued ribs adorn the menu. They open at 1700 hrs. and close at 2100 hrs. The price is a flat $9.95 plus tax. No credit cards but they will accept LOCAL checks.

Hot Springs (HOT) has the following approaches: ILS to runway 5, VOR, NDB, and GPS to runway 23 or 5. No approaches are available for 13-31. The airport is very well maintained and offers overnight tie downs for $3.00.

Lakeview (Gaston's - 3M0)

Gaston's Restaurant

Gaston's is a 3,200-ft. grass field located at Lakeview, which is near Mountain Home, on the White River. People come from all around to fish for the abundant trout. There is a beautiful restaurant, Gaston's Restaurant, which features the local trout. Lodging is available right next door at Gaston's White River Resort.

AMENDMENTS

My wife and I made the trip on a Saturday afternoon. It is about 2.5 hours from the DFW area. The runway was in pristine condition, as smooth as many paved ones. We taxied to park next to an old hangar and fuel pump. The restaurant is a 100-yard walk from the airplane parking area. Once inside we were seated in a big open dining room, built with large timbers, log cabin style. It hangs out just a bit over the White River. The popular tasty dish here is rainbow trout caught right out of the White River, which flows from Bull Shoals Lake. You can watch fishermen as they float down the fast-paced river, dragging their fishing poles. This is a great getaway for a day, weekend or week. There are small cabins available. You may also rent fishing boats and float fish all day. I know this isn't a hamburger place, but it sure is worth the trip. (501-431-5202)

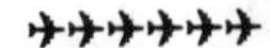

It is only a couple of miles from the infamous Whitewater Development.

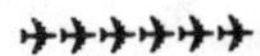

I concur with all the other reports. This is a great destination for the day, weekend or week. The grounds, scenery and river are pristine. The restaurant, a couple hundred feet from the tie down, is outstanding and the service is very friendly. We stayed overnight and ate lunch, dinner and the best Sunday brunch buffet I've had anywhere. Easy in on a great grass strip (runway 24) and out (runway 6). I didn't have a hamburger though!

Your reports about Gaston's are correct. It is probably the best place to fly-in for lunch in the U.S. The grass strip is extra smooth and plenty long; however, most folks who know the strip will land to the west and take off to the east, unless there is a very strong west wind. The field is in a valley with ridges on both sides. Power lines at the west end of the field could be an obstacle for those who choose to take off to the west on a hot day. They have fuel on the field. I recently met folks there who flew in from Houston just for a Sunday lunch (and Gaston's is in the extreme northern part of the state, only a few miles from Missouri).

The area around Gaston's also offers some great sight-seeing. The Buffalo River is a national scenic riverway. It joins the White River just about eight miles south of Gaston's. Two lakes are in the immediate vicinity and nearly visible from the traffic pattern.

Little Rock (Adam's Field - LIT)

Flight Deck

The Flight Deck at Adams Field (LIT) in Little Rock makes a great food stop. The restaurant is located inside the Central Aviation FBO building on the GA ramp at Adams. If you are refueling, you can grab a good meal while the avgas is being pumped.

The Flight Deck has a full restaurant menu and daily lunch specials. Yesterday's special was a small rib-eye steak with sides for $6.95, not bad pricing for an airport. Burgers and fries are also excellent. The fries are crisp, fresh, "skins on," just the way I like them.

The restaurant has plenty of seating space, good fast service, and overlooks the GA ramp. Lots of local folks drive to the restaurant to eat, so I guess it rates as one of the better spots around.

Morrilton (Petit Jean Park - MPJ)

Mather Lodge Restaurant 

No. 2 in Arkansas is Petit Jean State Park near Morrilton, Arkansas. They will pick you up from the airport for $3.00. The phone number is on the pay phone at the MPJ airport (501-727-5431). This place is as pristine as any in America. It is especially great for Father's Day or a very romantic Valentine's Day. There is a huge waterfall a short 45-minute hike from the State Park Lodge.

AMENDMENT

It is a nice place to drop in. The airstrip is huge, 6,000 ft. x 75 ft., given its light traffic volume and rural Ozark location. It was a mystery until I chatted with the park ranger who came to ferry us to the Lodge. Thousands of acres of land around about are owned by one of the Rockefeller Foundations. They built the field and donated it to the state, hence the impressive dimensions. It is worth noting the airfield sits high up on a bluff, 600 ft. above and south of the meandering Arkansas River valley. Taking off to the north is attention-grabbing on a hot day, rising terrain, covered solid with forest until you clear the ridge, and there suddenly is the Arkansas River. The effect is reminiscent of the opening sequence from The Sound of Music minus Julie Andrews.

You need to bring your own tie-down ropes. There are cables, but no ropes or chains at the field. I could not find the number at the pay phone to call for a ride, but I had copied the number from The $100 Hamburger so all was well. The Lodge is a splendid old rustic log edifice. Its entryway "frames" a gorgeous view to the west down a spectacular gorge. There are lodge rooms and cabins, some with kitchens, one with a hot tub.

That's the good news. The food at the Lodge was totally disappointing. It reminds me of the depressing Fred Harvey catering at the Grand Canyon. I ordered some fried mushrooms; their centers were still frozen when served. However, if you avoid anything pretentious on the menu, you will be OK. There's a fair salad bar, and should you be staying overnight, be aware that no wine or liquor is served (or allowed) in the park. If I give it three hamburgers, two would be for the view. Ask for a table by a window, which are the nonsmoking ones.

I bought fuel at Russellville (RUE) about ten miles NW of MPJ. I'd recommend it as a friendly field for a top-up when visiting MPJ.

Mount Ida (Bearce - 7M3)

Crystal Inn Hotel and Restaurant

This is a real pleasant place to visit. The airport is managed by Dr. Fowler, who goes out of his way to make you feel right at home. Adjacent to the field is The Crystal Inn Hotel and Restaurant. We had their breakfast buffet. For $3.75 you can eat enough to demand recalculations of your weight and balance.

Pine Bluff (Grider Field - PBF)

Airport Cafe

Pine Bluff is convenient. The Airport Cafe serves standard fare.

Springdale (Springdale Muni. - ASG)

Airport Cafe

Springdale has a cafe located inside the GA building. It has an outdoor "deck" where you can enjoy your meal and soak in all the sights, sounds and smells of an active airport. The GA building is a two-story modern facility. The cafe is located on the second level. Sadly, it is closed on the weekends.

Texarkana (Texarkana Regional - ASK)

Park Place Restaurant

There is a very fine restaurant in Texarkana called the Park Place. When you fly into Texarkana Regional (ASK), don't bother asking the FBO for directions. They didn't even know the place existed. The Park Place is owned by a very sweet lady from Vietnam. She specializes in French food. The filet mignon is without a doubt the best I have ever had. It's a bit pricey, but well worth it. Buy fuel, get the courtesy car and drive straight out along Arkansas Boulevard. The restaurant is less than a mile away on the right-hand side. There is a big sign that says "Park Place Restaurant." It's worth a try after a long day in the air. Enjoy!

Walnut Ridge (Walnut Ridge Muni. - ARG)

Airport Restaurant 

On our way back from Oshkosh, we stopped at Walnut Ridge (ARG) for fuel and a quick rest. The airport has a very small on-site restaurant. They serve burgers and sandwiches. The food was good. It turned out to be an excellent stop on this route.

Angwin (Parret Field - 2O3)

Pacific Union College Cafe

Fly into this slightly challenging hilltop Napa County airport, then walk a few blocks down the hill to Pacific Union College for a genuine Seventh Day Adventist vege-burger in their cafeteria.

Apple Valley (Apple Valley - APV)

Wings Cafe

Wings Cafe has the best pancakes I have ever had. APV often has no wind when WJF is blowing 20.

AMENDMENTS

Stopped by on our way to Alaska from Texas and had by far the greatest hamburgers California has to offer. Friendly people and great service. They will let you sign your name on the wall when you are finished. Ask Mike or Julie to take you up in their Extra 300, but be sure to do it before you head over to the Wings Cafe.

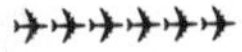

Nothing special food wise, simply average fare. The service is sometimes slow on weekends. This is a nice trip out of the LA basin and into the desert. On the way home you can do touch-and-goes on the huge ex-George AFB, now SoCal International, runways. They have no pesky controllers to bother you.

We had a funny experience there once. The wind had changed a little so we chose a different runway than normal. We were on final when we saw a red truck with flashing lights coming at us and heard "garble-garble airport security for runway check" on the radio. In the surprise of the moment, I thought I had done something very wrong and was being busted! As it turned out, he was just doing a FOD check. What a relief. We left in a big hurry anyway!

Atwater (Castle AFB - MER)

Castle Air Museum

My friends and I have made a food discovery. Castle Air Force Base, a recently closed Strategic Air Command base, has a museum with a restaurant. It is one of the best around and the food is even better.

Castle is located on the charts as Atwater/Castle.

AMENDMENTS

This field, no longer used by the USAF, has been opened for public use. There's currently one FBO. With any fuel purchase, you get two free passes to the Castle Air Museum, which is located on a different section of the field. It has an excellent set of aircraft, including an SR-71, a B-36, an Avro Vulcan, and even a CF-100 Canuck! The museum has its own snackbar, which is okay — nothing great, but on the other hand, it's reasonably priced — no $8.00 burgers.

If you're coming in on runway 13, don't practice your short-field landing technique unless you're ready to taxi almost 3 miles to the FBO!

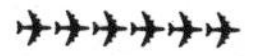

100LL is HIGH but it comes with free tickets and a free ride to the Air Force Museum, which is about a mile from the terminal. The museum is a great place to view unusual war birds. It is open every day except Thanksgiving, Christmas and New Years. The cafeteria is excellent and should take care of your hunger pangs.

Auburn (Auburn Muni. - AUN)

Wings Grill

The Wings Grill offers the standard fare. The FBO has relatively inexpensive fuel, so you can help the aircraft owner by tanking up.

AMENDMENTS

The Wings Grill is the only cafe at Auburn Municipal. It is open for breakfast and lunch seven days a week. They just did a remodel of the inside dining area. It looks good. You can also dine on the umbrella-shaded patio. This is a great place to watch and listen to the planes while having a meal.

The FBO's pump attendants will assist you with parking. It is crowded on the weekends.

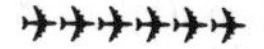

The Wings Grill added a latticed cover to the patio and was able to remove the umbrellas. Eating outside is now very comfortable. They have started dinner service from 4:00 PM to 9:00 PM. Their menu changes weekly.

Just returned from brunch at Wings Grill in Auburn, CA. They are now serving dinners but only Wednesday through Sunday. The tables are set with white linen, candles, and flowers. You can dine on the patio, weather permitting. The Fall in Auburn is excellent. The prices are a reasonable $12.95 down to $7.95 for entrees. They have an excellent chef. The menu is indeed changed weekly. The new screened patio even has misters for the hot days! Plans for next spring include adding a fountain to the patio.

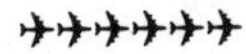

Wings Cafe has stopped serving dinner! They still do their wonderful breakfasts and lunches.

Auburn is a great place to do a day trip. Besides the good food at Wings Grill, you can see Old Auburn from the gold rush days and Central Auburn from the 1940s. Both are fun to visit. There are all kinds of nice shops and restaurants. If you want to stay longer, there are several good motels in town. On weekends there is even trolley service. It makes a circuit of the major motels, Old Town and Central Auburn for $1.00 — VERY good deal! You can rent a car through Wings Airport Service, so getting around is no problem.

Avalon (Catalina Island - AVX)

The Runway Cafe

I am somewhat of an airport cafe freak, having braved a crossing of the mighty Pacific Ocean just to get my oil-stained paws on one of those famous Buffalo Burgers at the Runway Cafe on Catalina Island, CA. There's nothin' like sitting out on the patio tasting those wonderful patties, all the while watching my meal's brothers and sisters roam the range.

AMENDMENTS

There is shuttle-bus service from the airport to all the restaurants in Avalon, so selections need not be limited to the burgers at the airport in the sky.

About a year ago, they instituted a pretty stiff landing fee: $5.00 PER PASSENGER. Personally, I don't mind because the airport is truly unique, and they don't get any funding from LA County or the Catalina Conservancy. However, some SoCal old-timers are burning up the pages of our local flying papers with protests.

This is my absolute favorite burger stop. The food is good. On a clear day, the view back toward the mainland is spectacular, and the ambiance, buffaloes and all, is neat.

Bakersfield (Bakersfield Muni. - BFL)

Airpark Restaurant 

Bakersfield has one of the best on-field restaurants in the state — moderately priced array of cuisine, mainly Italian/Californian, with some killer burgers and an unbeatable Rueben sandwich.

Of course, I don't know why you'd actually go to Bakersfield. It has a great big runway and a real friendly tower. The place is on the way to everywhere else, and there is always the country music. There's a cool motel at the airport next door to the restaurant. They love pilots and even let you bring beer from the restaurant to the hot tub and pool. There may be a more convenient place for an eat/stop/sleep-over, but I don't know where.

Bermuda Dunes (UDD)

Murphy's

The airport is about 13 miles east of Palm Springs, CA. It is right next to the 10 freeway and has a 5,002-ft. runway! From June-Sept. the temperature can soar to well above 100 degrees! Most of the private jets stop coming here during this time and use the longer strip at Palm Springs instead. About 100 yards south of the terminal you'll find Murphy's, a fried chicken institution! They offer many other menu items and have a nice bar adjacent to the restaurant. Most of the patrons are well into their 70s, but that goes with the territory.

Many celebrities use this airport because of its convenience to the posh Palm Desert area. During the peak season, November through April, the field is scattered with private jets! I have seen Arnold Palmer's Citation X here many, many times.

BE WARNED! Murphy's closes during the heat of summer!

Big Bear (Big Bear - L35)

Barnstormer 

My wife and I enjoy eating at Big Bear Lake's Barnstormer. The mountain air is clean, the people are friendly, the food is great, and the prices are perfect for those on a budget.

AMENDMENT

White table clothes, good food, and good service. They even have Chinese food. It is on the second floor, above the airport office.

Thelma's

Part of the adventure of flying, as we all know, is eating. I fly to Big Bear Lake about once a month. When it's socked in below, you can call (800-272-2967) for AWOS and usually fly in under clear and sunny skies.

Upon arrival, say hello to Marikay at the airport desk, walk one short block south to Big Bear Blvd., then one block east to Thelma's. This is biscuits and gravy at its best. They really have my favorite fly-in food, and lots of it, at reasonable prices. I really do like the place!

Bishop (Bishop Muni. - BIH)

Airport Cafe

If you have the airplane and the pilot skills to make this flight, Pete & Al's Airport Cafe has a pretty mean huevos rancheros, and most other breakfast items for that matter.

AMENDMENT

Pete & Al's is no longer in business at Bishop. The cafe has changed hands and become simply the Airport Cafe. The hours are 7:00 AM - 2:00 PM, Tuesday - Sunday. It's very basic menu that is not as good as Pete & Al's, but OK.

Bonnville (Bonnville - Q17)

Anderson Valley Brewing CO

There are a few good cafes and a wonderful restaurant or two in Bonnville. Don't overlook the Anderson Valley Brewing Company, a great place offering wonderful beer and tremendous burgers. It is my personal preference.

Town is only a 15-minute walk from the airport.

Borrego Springs (Borrego Valley - L08)

The Crosswind Cafe

Borrego Valley (L08) is in the desert south of Palm Springs. The beauty of the April desert wild-flowers is well known and makes this a great trip.

Spotting the strip can be a little tricky if your dead reckoning isn't up to snuff. When you do spot the place, you'll be surprised at the excellent condition of the blacktop. It is new.

The restaurant was closed for a couple of years but reopened in March '95. It provides the standard runway cafe menu. Get there before 4:00 PM if your ship needs fuel.

AMENDMENTS

Recently reopened and under new ownership. The specialty is the X-wind half-pound burger. There is also a quarter-pound version. You can buy an X-Wind Cafe coffee mug, cobalt blue with gold insignia, for $5.95.

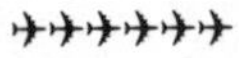

The cafe right on the airport reopened a few years ago and we have been a few times. This is a nice visit in any season but summer. It is bloody HOT then!

A little nicer than the average lunch spot. I think they do a nice dinner, too. Very good menu, Cobb salad, good sandwiches, etc., served on linen tablecloths. Prices are a little high but worth it.

The only less-than-favorable comment I might have would be about the "aire." The staff was not really friendly, to us anyway, kind of like the rich retiree in the nice threads who arrived in the Benz got the big smiles and the doting service and we, "you dare to appear at our door in tee shirts and jeans? We'll serve you if we must" pilot people got less. Our Cobb salad popped up at the kitchen window only to be given to another customer — type A, see above — who ordered some 10 minutes after us.

Anyway, it is a nice place to take a new girlfriend. Just dress expensively and wear a gold Rolex.

I'll go eat fries at Flabob!

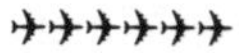

The restaurant is average, but you can sit up on the roof and if anyone's practicing in the aerobatic box just north of the field, which they will be on most weekends, you get a great airshow!

TRIP BONUS — Anza-Borrego Desert State Park Sky Trail

The California State Parks Department developed an aerial tour in booklet form to give pilots and passengers a chance to enjoy the natural and historic features of Anza-Borrego Desert State Park, some of which can only be seen from the air.

The booklet is in two (separate) parts: a navigation guide that is used by the pilot and a scenic explanation which may be used by the passengers. It was designed for one passenger to read aloud to the others.

The sky tour begins at the Borrego Valley Airport, traverses a 150-mile roundabout tour of the desert, and returns to the airport.

The navigation section is intended for VFR use, with headings, ground reference points, and optional GPS coordinates. The description section contains a little geography, biology and history of the local area. The hand-drawn illustrations are adequate for reference and do not detract from the vistas to be seen out the window.

The Sky Trail is well thought out, informative, and entertaining — an excellent way to introduce people to flying, and to the desert. Not a bad way to spend an afternoon, either.

The booklet is economically priced at $2.50 and should be available at the airport, but rather than take the chance that they are out, you may order one from:

"Anza-Borrego Sky Trail"
California State Parks
Colorado Desert District Office
200 W. Palm Canyon Dr.
Borrego Springs, CA 92004
Phone: 619-767-5311

Bridgeport (Bryant Field - O57)

The Bridgeport Inn

The airport is roughly equidistant between South Lake Tahoe and Mono Lake. The scenery and terrain are wonderful, if a little challenging. The airport elevation is 6,500 feet, and it is surrounded by mountains.

The Bridgeport Inn is quite nice and slightly formal, in a Western gold-rush town way. The Victorian & Fries is excellent, and yes, they do have hamburgers. It is a two-minute walk from the airport.

Hays Street Cafe

The Hays Street Cafe is more like a diner, although the menu is varied. It is an easy walk from the tie-downs. Really, this place deserves more traffic, although not during the winter, when most of the town is snowed in.

Bighorn (Byron Muni. - C83)

The Byron Inn

The Byron Inn is located about two miles from C83. Turn right off the access road and left at the stop. It's just over the tracks along the Byron Bethany Highway.

The food is great but I don't know how you could get there except walk. They are open from 5:30 AM until 2:30 PM and serve only homemade dishes. It was voted best breakfast in the county, probably for the terrific biscuits & gravy.

California City (California City - L71)

Da Vinci's

California City is north of Mojave almost to the southern tip of the Sierras, but still on the desert floor. The airport has skydivers and gliders. The record for mountain wave gliding was made from Cal City. Watch for unannounced traffic. Skydivers land southwest of the field. Gliders land parallel to the main runway, next to the taxiway.

The on-field restaurant is Da Vinci's. Terrific Italian food is offered at very fair prices. Five of us have eaten there on several occasions and sampled just about every dish. All were good. Some were excellent. The spumoni ice cream cake is worth the trip.

The Ready Room

For those of you with memories of the original Ready Room that burned down many years ago, rejoice! The "new" Ready Room is now open for breakfast, lunch and dinner, Tuesday through Sunday. The prices are right, the food is good and you can watch planes, skydivers and gliders while you eat. It's definitely worth a stop next time you're out this way.

Camarillo (Camarillo - CMA)

The Way Point Cafe

My father is a bigger nut than I am about flying and recently finished building his RV-4. He is based near the east end of Camarillo Airport in California and haunts the on-field burger joint.

The Way Point Cafe has a famous "Tri-Tip" steak sandwich. You can have it served on their comfortable patio. Signed pix of famous and not-so-famous aviators cover the walls.

AMENDMENT

Great weather 350 days a year! I had the Tri-Tip lunch with beans and potato salad out on the patio.

Cameron Park (Cameron Airpark - O61)

The Breakfast Nook

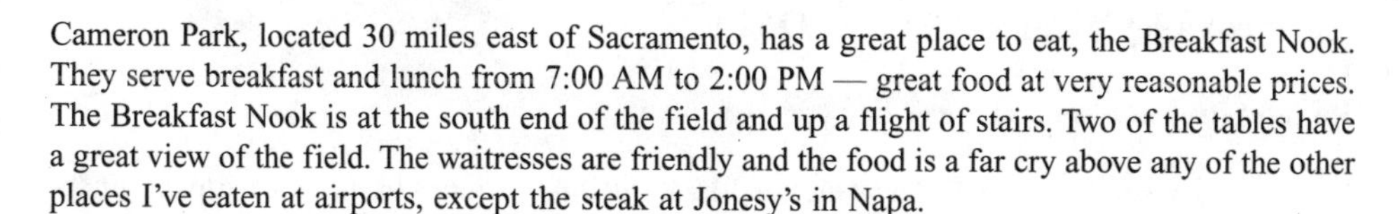

Cameron Park, located 30 miles east of Sacramento, has a great place to eat, the Breakfast Nook. They serve breakfast and lunch from 7:00 AM to 2:00 PM — great food at very reasonable prices. The Breakfast Nook is at the south end of the field and up a flight of stairs. Two of the tables have a great view of the field. The waitresses are friendly and the food is a far cry above any of the other places I've eaten at airports, except the steak at Jonesy's in Napa.

AMENDMENT

The Breakfast Nook has moved from the Cameron Park Airpark to Burke's Junction, which is about 1.5 miles south of the field. It's still a great place to get breakfast, or lunch for that matter, but it is no longer located on the field. There are no restaurants located on the airfield at O61, but a mile and a half isn't so far to walk.

Carlsbad (McClellan-Palomar - CRQ)

The Olympic Resort

McClellan-Palomar airport in Carlsbad, CA (CRQ), has traditionally had a pretty good cafe. It's on the second floor of the pilot's shop with an outdoor terrace and a good view of the action. It closed last fall but was supposed to reopen under new management in Feb. Better call ahead to be sure. If they're still closed, walk across the road to the Olympic Resort. It is a great spa open to the public with a really good restaurant.

The airport is in a very scenic, sunny setting just inland from the coast. Unlike the other San Diego-area fields, you don't have to mess with the Class B airspace if you're coming from LA or points north.

Airport Cafe

I fly to Palomar once a week; the food is great! I believe the cafe is open breakfast through 6:00 PM. If you arrive and they are closed, my kids like the microwave pizza in the Cinema

Air pilot's lounge just down the field. Don't miss a walk through the private museum in their hangar. You'll see perfectly restored and maintained warbirds, super cubs and other Hollywood movie airplanes.

AMENDMENT

Palomar offers a big menu and good service. You can sit out on the balcony and watch the planes. This is a very busy spot on weekends, so brush up on your controlled airport manners. We once arrived just in time to see a formation of a P-51 and a P-47 takeoff.

There is a female controller here who is renowned far and wide for her skill at keeping 14 planes all going the right way at the same time and never missing a beat. She does it all with a smile. Yes, you can hear a smile over the radio. This alone is worth the trip.

The Cinema Air hangar is fabulous. They'll let you wander all around alone among their many, very expensive planes. A beautiful collection of warbirds that look as if they were made today. Don't miss this 'port!

Carmel Valley (Carmel Valley - O62)

Running Iron Restaurant

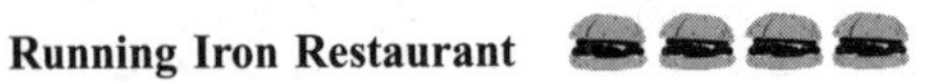

We recently visited Carmel Valley (O62) and had a delightful lunch sitting on the patio of the Running Iron Restaurant. Follow the path to the post office on the south side of the field and continue for one block. The restaurant is on the right.

Be advised that there are no services at Carmel Valley, and no night operations. The runway is short (1,800 feet) and the surface is rough.

Although I've never been charged a landing fee, I have heard reports that the airport is now extracting a $10.00 landing fee for nonresident aircraft, except for antiques dating from before 1928.

AMENDMENT

I recently flew in from Reid-Hillview. It is just south of San Jose International. I stayed east to avoid the Monterey Class C, then turned west down the valley for a straight in. Beautiful ride!

Just a short walk down from the Running Iron Horse is a coffee house and deli that sells Lapert's ice cream for a dollar a cone! Maybe that's really a hundred and one dollars a cone, depending on your point of view. Regardless, great ice cream! Look for the Lapert's sign hanging in front of the store.

We flew home through Monterey Class C and up the coast. Monterey is not too busy and provides great Class C practice for the timid.

Chino (Chino - CNO)

Flo's Cafe

A legendary eating place in the Southern California area, near Los Angeles, is Flo's Cafe, at the Chino (CNO) airport. Flo's is an active meeting place for Sunday breakfast for a large number of pilots, although it is busy nearly anytime. Their biscuits and gravy is the most popular breakfast dish.

A big attraction at Chino is the large number of warbirds on the field. There is a museum where you can see some really rare planes for a small admission charge. You can also see many for free just parked around the field.

Chino is also a popular field for aircraft builders/pilots to fly off the first 40 hours on their experimental aircraft. The FAA considers it to be an uncongested area. It is the only airport with a long runway and emergency equipment in that category in Southern California.

Be aware that Chino underlies Class C airspace so it is best to contact Approach and get a transponder code before entering. This can be a busy area! When you land at CNO, tell ground control you want to taxi to Flo's and they will take it from there.

AMENDMENT

Home of the best and most cholesterol-laden biscuits 'n gravy, and two excellent aviation museums. Lots of interesting planes are based here.

Chiriaco (Chiriaco Summit - L77)

The Route 66 Cafe 

Chiriaco Summit is just north of the Salton Sea and is the remnant of an air field that served Gen. Patton's Desert Training Center. A five-minute walk towards the freeway (I-10) transports you back in time to a cafe from the Route 66 era. The burgers are straight Americana, and you can wash them down with an unusual but tasty date shake.

To top it off, there is a museum 100 paces further west dedicated to Ol' Blood and Guts himself. About a dozen WWII and later era tanks surround this large museum. It contains displays from the Civil War through Desert Storm.

I bought a dud grenade there and tossed it unpinned onto the conference room table during a tense moment at work. A real icebreaker!

Also, there is a plaster topographical model of the Southern California, western Arizona, and southern Nevada area. It was used in the 1920s by the architects of the aqueduct system to plan the dams and pumping stations that distribute water from the Colorado River to Southern California. This thing is huge, weighs tons and is to scale!

AMENDMENT

Fewer airport people and more "I've been driving for hours and I'll eat anything just to get out of the car with those kids!" folks. It is well worth the flight to see the Patton Museum.

Coalinga (Harris Ranch - 3O8)

Harris Ranch Restaurant 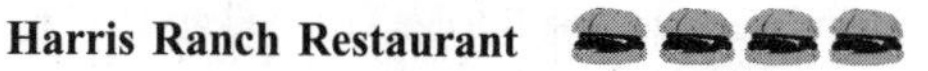

The Harris Ranch near Coalinga is well known and popular. Their strip is 2,800 feet of asphalt. If you become tired after dinner, spend the night at the Inn at Harris Ranch!

AMENDMENTS

Harris Ranch provides excellent beef. The runway is a little tight for most light twins, so some careful planning is in order if you plan to feed a plane-load of people.

They sell their beef as well as serve it. You can take some home vacuum packed and cooled in an ice container. If you choose to stay, the hotel accommodations are moderately priced ($$) for above average rooms (***).

This is one of the few aviation destinations you can find via the olfactory senses. The Harris Ranch Restaurant is very close to a large cattle operation. It is about 200 miles southeast of San Francisco, almost exactly between SF and LA.

They have a nice steak selection and a good hotel but are relatively high-priced. The food is good but not great. Beef lovers will rejoice! Check out the Rocky Mountain Oysters. Service is marginal, as you might expect from a Highway 5 motorist trap. The airfield is right on the property. The 29-foot-wide runway works up the old appetite.

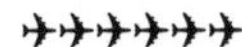

Harris Ranch airstrip has a new top-coat of black-top, and new markings as of 2/1/97. The tie-down area has been swept clean of loose gravel. The wind is usually 10-15 knots but straight down the runway most days.

After lunch we bought some take-home beef, which the clerk vacuum packed and wrapped with a special ice-pack they supplied at no extra charge. Wow! Those were just about the best steaks I ever had, anywhere!

Columbia (Columbia Muni. - O22)

City Hotel Restaurant 

Columbia, for those in the know, is one of the TOP TEN fly-in eating places in California. Here's why.

The walk into town from the airport is through a nature trail in a forest and takes only 15 minutes. Once in historic Old Town, you'll find the City Hotel, which has a restaurant that also serves as a culinary academy in cahoots with the local community college. It's the City Hotel Restaurant; go figure. The menu is strictly first-rate, offering exceptionally fine cuisine. Features are mostly French-chef style with incomparable presentation. Meals come out looking like works of art. The service is the very best, and the wine list is impressive.

In the off season you can sometimes get in without reservations, but I'd recommend phoning ahead. Why risk it? Prices will be around $40 per person, including wine. You'll leave satisfied and feeling that you have found a GREAT bargain!

Castello's

There is a new buffet place called Castello's. It has good food and is only a 10-minute walk from the airport. It's well worth the flight.

Columbia House Restaurant 

That Columbia is one of the top fly-in eating places in California is evidenced by the number of aircraft taking off and landing there every weekend!

One more recommendation for breakfast: the Columbia House Restaurant on Main Street is a good value, with the nice touch of waiters and waitresses in gold-rush-era costumes.

TRIP BONUS Columbia Airport Campgrounds

Where's Columbia, THE MECCA for northern California fly-ins? There's no restaurant on the field but a great campground where you can cook 'em your way! If you prefer, it's only a short walk to historic Columbia. This destination is standard fare for new northern California pilots.

Concord (Concord - CCR)

Atrium Court 

Personally, I like going to the Atrium Court Restaurant inside the Sheraton at Concord. Given the usual landing direction, you'll taxi all the way down to the end of the runway and then hang a right. It's a nice place for lunch or dinner, with tablecloths, trees, and a running brook inside, also very reasonable prices. If you want something more Las Vegas-ish, with a lot of variety, the Peppermill is right across the street.

AMENDMENT

No matter which runway you arrive on, it would not be a right turn into transient parking, unless of course you landed long on runway one, then turned right at the end and taxied the mile back to the approach end. Actually, the directions wouldn't be too bad if you landed on runway 19 and turned left. However, runway 32 is the normal runway for takeoffs and landings, even when LDA 19 is the IFR runway.

The Peppermill

The Peppermill is one of a chain of high-end coffee shops found around northern California and is good in its own way.

The easiest method for newcomers to find the restaurants is to request progressive taxi to the Sheraton. Transient parking is right next to the hotel.

Corona (Corona Muni. - AJO)

Bob's Cafe 

I just tried Bob's Cafe in Corona, CA. My friend swears that it is the absolute best cafe he has ever flown to. I enjoyed it but can't swear to its Ultimate Burger standing without a few more visits. The chili burger was tremendous! They even make their own potato chips.

AMENDMENT

Good food, the usual stuff, lots of pilots hanging around. Known to locals as the "food simulator," but I think they are just kidding. It's less frequented by my crowd 'cause they can't burn any 80/87 getting there.

Death Valley (Furnace Creek - L06)

Furnace Creek Inn 

There are two airports in Death Valley National Park. The largest is at Furnace Creek. The other is at Stovepipe Wells. The runway at Furnace Creek (L06) is small and not terribly well maintained, 3,065 feet x 70 feet and 211 feet BELOW sea level!

Furnace Creek is the location of the Park Service Visitor Center and the Borax Museum. You can see the relics of the 20-mule-team days.

Furnace Creek has the best restaurants, including the Furnace Creek Inn. The Inn is about a two-mile walk from the airport, or you can call for a shuttle van. Best time to visit is spring and fall. Avoid summer, as it can be very hot. They don't call it Death Valley for nothing!

The Furnace Creek Ranch 

The Furnace Creek Ranch has several less expensive places to eat, including its Steak House, a coffee shop and the golf course sandwich shop.

Death Valley (Stovepipe Wells - L09)

Stovepipe Wells Motel & Restaurant 

Stovepipe Wells has only one dining spot, the motel's restaurant. It is only a quarter-mile walk from the airport. The food is just OK, but the views are great to spectacular!

Stovepipe Wells (L09) is 3,265 feet x 65 feet and less well maintained than Furnace Creek. It is 25 feet above sea level.

Fresno (Chandler Downtown Airport - FCH)

Flying Saucer Restaurant 

The Flying Saucer Restaurant, located in the historic terminal building at Chandler Downtown Airport in Fresno, California, features Chinese, Mexican and American dishes. All are good. It is open 7:00 AM to 3:00 PM, seven days a week. This charming early-thirtys facility has large windows facing the ramp for perfect aircraft watching.

Fullerton (Fullerton Muni.- FUL)

Airport Cafe

A very good restaurant is in the terminal building. It is very busy at lunch. The service is quick and the food is good.

Georgetown (Georgetown - Q61)

Skyway's Sandwich Shop

Georgetown sits in the middle of the California gold country. On the field, you'll find a sandwich shop at Black Sheep Aviation. If that doesn't ring your bell, there are several restaurants in town. You can also camp out here. Call the Campground manger at 916-333-0810 for details.

If you are new to mountain flying, be careful. It can be full of surprises!

Fly safe, plan ahead, but most of all enjoy this extraordinary airport. In the travel industry, I think it would qualify to be called a DESTINATION!

AMENDMENT

I just returned this morning from a personal viewing of the Georgetown, CA, airport. The picnic area and campgrounds are threatened by logging its trees to satisfy county interests. They were selectively cut so that the campground and picnic area are still usable. It is quite a bit sunnier than it was before.

The West Coast Cessna 120/140 Club just held its fall gathering at Georgetown and pronounced the experience satisfactory, but many of the club members did sign a letter to the county supervisors urging them to spare the recreational area from more damage. I joined in the signing. The cafe at Black Sheep Aviation is open and as friendly as ever. Stop by next time you are in the vicinity and see for yourself.

Grass Valley (Nevada County - O17)

The Warehouse Deli

There are no eating places on the airport but try the Warehouse Deli. It's only a 1/2 mile walk. From the administration building, go down the hill and turn right. Ten minutes later you will see it on your left. Order your sandwiches at the counter. They will be delivered to your table.

AMENDMENT

Oscar 17 is adjacent to two of the nicest towns in the Mother Lode, Grass Valley and Nevada City. If you want more than the Deli can provide and have the time, go into town. The cab ride is $10.00 each way. If you are lucky, the caretaker on the field will have his classic 1963 Chrysler available for rent, half a day for $25.00 and all day for $40.00.

Grass Valley is a beautiful Victorian town with lots of shops and a few good restaurants. The Empire Mine State Park and Mining Museum and Malakoff Diggings State Park are all within a short distance.

There is a campground on the field with picnic tables, running water and flush toilets. At this time there is no fee for the use of the campground. The scenery is absolutely beautiful!

Gravely Valley (Gravely Valley - 1Q5)

Lake Pillsbury 

Indulge your Alaskan bush pilot fantasy and land on this public UNPAVED strip on the shores of Lake Pillsbury. Bring your own barbecue supplies and food to cook. If you like, camp right off the runway in the fly-in camping area. No running water or other facilities, but the campgrounds are within walking distance. A marina with a store is a few miles easy level walk around the lake.

Groveland (Pine Mt. Lake - Q68)

Corsair Coffee Shop 

Pine Mountain Lake (Q68) is a public airport owned by the county of Tuolumne. I frequent the Corsair as we have a weekend cabin at Pine Mountain Lake. Heidi and Marv are the proprietors. Both are pilots. Monday and Tuesday they close to go flying. Breakfast is great with the best $100 pancakes you can buy. They also serve lunch and dinner.

AMENDMENTS

Marv and Heidi are now open on Monday and closed Tuesday and Wednesday.

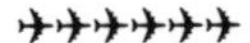

I flew in for lunch. This is a Super airport. The Corsair Coffee Shop has good food and fast service! Definitely a great fly-in.

Half Moon Bay (Half Moon Bay - HAF)

The Fish Trap 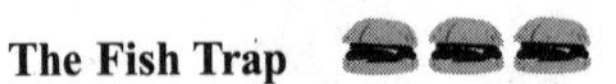

My favorite place is Half Moon Bay. Quaint shops and restaurants are all within walking distance of the airport. Park and tie down at the SE end of the runway. Barbara's Fish Trap is a very easy walk. It offers good fresh fish.

AMENDMENT

Hey, don't underestimate Barbara's Fish Trap! We fly there several times a year and much prefer it to the other choices. Yes, it's quite informal, but the seafood is always fresh, well prepared, affordable and worth the 15 minutes or so you usually have to wait in line.

The Princeton Inn has reopened with a nice looking dining room and menu. I didn't try them out because the menu is short on seafood, but plan to have a go soon. Looks like a potential dinner spot. I think they have rooms also but didn't inquire. You can't miss the place. It's the first habitation on your right when you take the shortcut path from the south tie-down area.

The Shore Bird

Oh, my goodness! You don't do Half Moon Bay, just SW of the city of San Francisco, justice! The Shore Bird is quite good for seafood cuisine, perhaps even above average. The Fish Trap is more the quick-and-dirty type of seafood restaurant.

AMENDMENTS

I flew my Cessna to Half Moon Bay (HAF) and had a great brunch at the Shore Bird restaurant. It opens at 11:30 AM. Heads up, as it is a very busy little airport with right downwind traffic.

Nearby, Castle Air Base is now open to private pilots and has a nice museum.

✈✈✈✈✈✈✈

Castle Air Force Base, or whatever it's called now, is not really "nearby." It's 1-1.5 hours away, over in California's Central Valley, south of Modesto.

I concur with the above reviews. The nearby town of Princeton-by-the-Sea was described by a Bostonian friend with whom I had lunch as "very much like Cape Cod." Try to get one of the outdoor tables at the Shore Bird.

The best place to park your plane is the transient area, down south of the approach end of runway 30. There are usually plenty of tie downs. Once a month, a number of local law enforcement agencies cordon off the route to that parking area and all the nearby taxiways to do a bit of high-pursuit driving practice. It's actually a bit fun to watch. What's more, you can park your plane short of the cordon and someone will readily give you a ride across the temporary Starsky & Hutch practice area to the footbridge into Princeton. On your way back from lunch, stand at the open area near the bridge, away from the cars and screeching tires. Someone will pick you up and drop you off at your plane. Free taxi...no kidding!

✈✈✈✈✈✈✈

I just learned that the high-speed driving training occurs on the first Wednesday and Thursday of each month. When the training is in session, it's best to park the airplane in the semicircular area marked at the end of the hangars at the southeast end of the field.

✈✈✈✈✈✈✈

I had lunch at the Shore Bird Restaurant on the south side of the airfield. Actually, it's in Princeton-by-the-Sea. My friend had been there about a year or so ago and wanted to return. The lunch menu has only a few items under $12.00! I tried a bowl of seafood chowder for $8.00. My friend had a shrimp and something sandwich. The bill plus tip was $30.00. We shall try the other places recommended in the preceding reviews.

Ketch Joanne's

There's another great restaurant at Half Moon Bay (HAF) called Ketch Joanne's. We usually go for breakfast, which is always delicious. The walk to and from the south end of the airport across the bridge is pleasant and a nice stretch after a long flight or a big meal. Watch out for the poison oak along the creek!

30 Cafe

The on-field restaurant, now named 30 Cafe has changed ownership. You will instantly know that more than the ownership has changed, because the place is crowded. You have to sign a list, but the wait is reasonable. Breakfast is quite good, prices are fair, but the service may be a bit spotty. So try eating on the field while you watch the planes, then walk to the beach with a full tummy!

Hawthorne (Hawthrone Muni. - HHR)

Icarius Cafe

Hawthorne (HHR) is very close to LAX. Getting in can be intimidating if you aren't used to sharing airspace with the heavy metal. The cafe overlooks the runway and the flightpath to LAX. The food is pretty standard for any airport.

Recently, a light rail system has been constructed in Los Angeles. There is a stop very near the airport. This gives you pretty decent access to downtown, the South Bay beaches, or East LA to see the famed Watts Towers. This airport is adjacent to Northrop. You'll see some experimental aircraft roll out once in a while.

Hesperia (Hesperia Air Lodge - L26)

Airport Cafe

Hesperia's airport is a few miles south of Apple Valley. The cafe here is nice. This used to be a favorite of my bunch for a Sunday trip. They serve a great breakfast.

Hollister (Hollister Muni. - 3O7)

Ding-a-Ling Cafe

The Ding-a-Ling Cafe is a great stop if you want a very good breakfast or lunch. Open every day.

AMENDMENT

The previous review is still valid. The Ding-a-Ling Cafe is a real bargain and has friendly help.

Jackson (Westover - O70)

The Oak Tree

The Oak Tree restaurant is actually in Sutter Creek, a short walk from Westover Field, about two blocks north on Highway 49. It doesn't look like much from the outside, but I find it quite satisfactory. They have a pretty decent menu selection. All items are good, and the servings generous. It's not a place you'd take a visitor you wanted to impress, but it is more than adequate for a hungry pilot.

AMENDMENT

A short walk to average food, average service, average ambiance, and average prices. I'd give it a two burgers rating.

Kern County (Kern County - L05)

Pine Cone Inn

Land at Kern Valley Airport, next to Lake Isabella. Walk into town and eat at the Pine Cone Inn. Say hello to Roberta. It's really a wonderful place to stop.

Airport Cafe

This is one of the best out of the way mountain airports around — great little cafe on the field, nice campground next to the runway and Kern River fishing just 50 yards away.

Kernville is a short walk away with good inns and restaurants.

Lakeport (Lampson/Lake County - 1O2)

Skyroom Restaurant

I like to run up to Lampson Field at Clear Lake and eat on the field at the Skyroom Restaurant. It's a nice scenic flight followed by decent food.

AMENDMENTS

The restaurant at the Lakeport, CA, (Lake County) airport is right next to the runway, situated so you can watch the planes come in and take off. It has a great Sunday brunch and really good food the rest of the week.

Lake County is north of the Sonoma and Napa County vineyards.

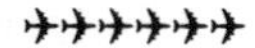

I flew up to Lampson on a Monday recently, mid-afternoon, and was told "we only serve sandwiches and hamburgers this time of day." I ordered a double bacon burger — not on the menu, at least the "double" part, but they did it anyway. I got a couple of slightly charred patties, a huge mound of dry bacon, and more onions than I could shake a stick at. Should I say, "at which I could shake a stick"?

The french fries were excellent. I asked for and received a full pitcher of iced tea.

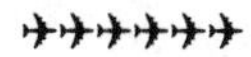

They must have made some changes. We found the Skyroom Restaurant really great for lunch. Try the shrimp salad. It's huge and super good. I'll stop again.

Lake Tahoe (South Lake Tahoe - TVL)

Scusa's On Ski Run

PIREP

Our personal favorite in South Lake Tahoe is Scusa's On Ski Run, a pricier Italian restaurant located a few miles from the airport on Ski Run Blvd. Expect to pay over $80.00 per couple, including appetizer, salad, entree, dessert, and wine. Don't forget to budget another $20.00 for the cab ride!

Tahoe can be a treacherous airport to land at or take off from due to its high elevation and prevailing winds.

Pizza Joint

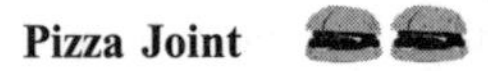

If you want to turn that burger into a $100 pizza, there's a good pizza joint about a 15ish-minute walk toward town from South Lake Tahoe Airport. Works for me!

Lancaster (General Fox Field - WJF)

Sly Fox

General Fox Field (WJF) is in the middle of the Palmdale Desert, not far from Edward's AFB. I'm always surprised that this is a towered field until one of the neighborhood jets decides to stop by for a grease burger. I have been there when two F-14s landed and the pilots wandered in for a patty melt at the local cafe. The desert is often EXTREMELY windy, so be careful to read the forecast before coming in.

AMENDMENTS

Sly Fox cafe is still operating with a regular weekend breakfast clientele. I am unqualified to rate it because, as my wife says, I will eat a greasy shop rag if I get it at an airport. Their food is good enough, and they have some pretty good "home cooked" desserts.

Apollo Park is a 10-minute walk away from transient parking for a nice picnic area.

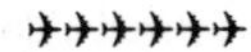

The field is home to the U.S. Forest Service's Fox Tanker Base, located on the east side. This base is designed to accommodate large numbers of airtankers, aircraft that drop aerial fire retardant, as well as Leadplanes, Beech Baron 58Ps that lead the airtankers over the fires and air attack planes, usually Aero Commanders or 0-2s. The base is manned year-round but is usually busy from May to November.

We are in the process of adding new facilities to the base, and one day hope to have a viewing area for the general public. I'm told this is still a few years off, due to budget constraints. During large fires, we usually have a few people outside the fence, where they can get a good look at our DC-4s, DC-6's, DC-7s, C-130s, P-3s, PB4Ys, S-2s, P2Vs, and SP2Hs.

Please be careful around the field during a major fire and yield to our aircraft, as seconds can make the difference for someone's house. During slow times, give us a call and we will be happy to show you around. Call (805-948-6082) and ask for Martin or any of the Forest Service pilots.

La Verne (Brackett Field - POC)

Norm's Hangar 

We recently visited Norm's Hangar Restaurant. The building and atmosphere are very nice. There is ample seating and a nice view of the AOA.

I must say, however, that I was quite disappointed with our entire dining experience. First of all, the service was very slow! We were there at the lunch hour; however, there were many waitstaff on duty. Nobody came to our table for about 15 minutes. Then they got our order wrong. The food was OK, the usual airport cafe fair.

When I went to pay, I found out that they DO NOT ACCEPT CREDIT CARDS! I had to go out to the plane and scrounge up all the loose change from my flight bag. To make matters worse, I went for an after-lunch mint and was informed that they charge 15 cents for each one!

It will be a long time before I return to Brackett for lunch!

Little River (Little River - O48)

Cafe Beaujolais 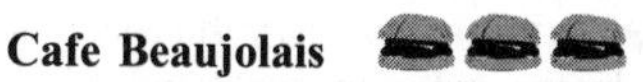

You'll need to rent a car from the FBO. Call Coast Flyers at 707-937-1224. Drive into Mendocino, about 15 minutes away. Do some sightseeing and shopping in this quaint New England-style town. Then have breakfast, lunch, or dinner at the Cafe Beaujolais. Act like a local or a regular and call it the Cafe B. I don't have the number; call directory assistance for Mendocino and make a reservation.

AMENDMENT

Leaving OAK in my PA28-140 and following the coast northwest brought me to Oscar 48, Little River, in about an hour and a half. The runway is clearly visible for 10-plus miles and is long and wide, possibly for fire fighting tankers. Field elevation is 527 feet, on the day I went it stayed above the stratus layer that periodically covers Mendocino proper.

There are two ways into town. Use the Mendocino Stage, available on call at $8.00 per person, or rent a car from Coast Flyers for $30.00 to $40.00 per day, depending on the size of car, expected mileage, etc.

I chose the rental route as I didn't know the proximity of things and my 6-year-old was cranky enough already. As it turns out, with fewer than four people, it's probably more cost efficient to use the shuttle van, assuming time isn't an issue. Everything in town is within easy walking distance.

My wife, a shopaholic, was in near heaven. There are many dozens of shops on virtually every street. The bulk of them offer the standard tourist fare of t-shirts and "Mendocino" labeled knick-knacks. Some have very nice jewelry, home and garden decorations, clothes, etc. Food at most restaurants is reasonably priced, but watch out for the $12.00-plus entrees. We only stayed for the day, partly due to impending coastal stratus, partly due to the aforementioned cranky one.

My plane partner has stayed here before with his ladylove and reported that very pricey to reasonable bed and breakfasts abound. The coastline is as dramatic as they come. Make sure you bring warm clothes as the temps here drop rapidly in the evening.

Livermore (Livermore Muni. - LVK)

Cattlemen's 

Out in Livermore, the Red Baron is long gone, having been burned to the ground in a fire. Its parking area is replaced by things that service the golf course. Fortunately, there is a Cattlemen's within a very short walk of the transient parking area, right at the entrance to the airport. The Cattlemen's chain has always been great for steaks, ribs, etc.

AMENDMENT

Having not been to other Cattlemen's, I can't compare. However, I've eaten at the LVK one a couple of times and found it acceptable. Be prepared for a wait. You can hang out in the bar/lounge, and order your drinks. No ethanol for the pilot, 8 hours bottle-to-throttle! Ask to transfer the tab to your meal check.

Many meals come with salad — not your run-of-the-mill little saucer of lettuce and carrot shavings. They bring out a big bowl and toss a number of excellent salad fixings, then divvy the contents into each diner's bowl. It's unlikely the mixing bowl will be empty by the end of the meal, unless all the diners are metabolic infernos.

Beebe's Sports Bar and Grill 

Taxi to the west end of the airport and go to Beebe's Grill. Hamburgers, burritos, etc. all are pretty good and reasonably priced. It's located on the golf course. Why not play a round?

AMENDMENT

We flew in the other day and tried Beebe's Sports Bar and Grill on the NW side of the airfield at the golf course. Good food and quick service. Worth another trip!

Lodi (Lodi - 1O3)

Lodi Airport Cafe 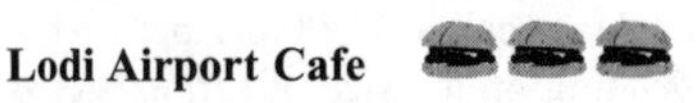

When the parachute jump zone is operating, this is a pretty neat place to hang out...especially when the DC-3 is flying. The Airport Cafe has standard run-of-the-mill food, reasonably priced.

There's a VOR approach, but if you need that, the jumpers won't be there. The runways are pretty narrow, which is a good way of convincing yourself to land on centerline.

AMENDMENTS

The Airport Cafe is OK. It's located at the end of runway 26. It's open every day, weekdays 7:00 AM to 2:00 PM and weekends 7:00 AM to 3:00 PM.

This is exactly what an airport burger place should be. It's very neat and clean and well maintained, rather out of character with the rest of the airport.

Caution! Runway 08-26 is listed as 42 feet wide and is pretty rough in places. On departure from runway 26, watch for jump planes, likely to be departing on runway 30 regardless of wind direction. Also be alert for ultralights. Runway 26 is right traffic to avoid the drop zone.

A must-see in warm weather — really good burgers and fries and the service by the gals keeps 'ya coming back!

Lone Pine (O26)

The Smoke Signals

Halfway up the Owen's valley is a town called Lone Pine. It is between Bishop and Inyokern about 10 miles north of Owen's Dry Lake. The airport has runways that are long, hard and lighted. I can't believe anyone hasn't written about the restaurants in this town and one in particular, The Smoke Signals. After a long day flying or crewing gliders, I always stop in. They are seasonal and open only in the summertime. The barbecued ribs and meat are smoked by the owner and defy description. They are cooked on a cobblestone grill and stack he built himself.

It is patio dining. The decor is rustic, like the Old West. Lanterns, barrels, and fencewood make up the furniture. Both myself and guys I know have flown Cessnas up from Los Angeles just to eat there. Hint: Make a reservation and then walk the street and see the shops. Be careful flying in and stay west of 395. You don't want to get in the way of the low altitude training route or the cruise missile test range.

Los Angeles (Whiteman - WHP)

Crosswinds Cafe 

Recently, we visited the Whiteman Airport, located in the valley north of the Burbank/Glendale/Pasadena Airport.

We stopped in for a hamburger at the Crosswinds Cafe. The tower was very helpful in locating transient parking for us. The facilities outside were clean and simple. You enter the restaurant through the bar, which stays open until 2:00 AM and seems to have a regular locals-only crowd.

The main part of the restaurant was closed off at the time we visited, so we ate in the bar, alongside the locals. The menu is basic breakfast and lunch, with hamburgers, sandwiches, soup, and salad. What got my attention right away was the price — VERY CHEAP! Imagine, $2.75 for a cheeseburger and fries. The food was great. The hamburgers were good and the french fries were done perfectly, not greasy or burnt. The staff was friendly, so were the locals. Thumbs up for Whiteman Airport and the Crosswinds Cafe.

Los Banos (Los Banos Muni. - LSN)

McDonald's

The true $100 hamburger! Los Banos, California, is right across the highway from a McDonald's. It works if you're hungry.

Ryan's Restaurant

Ryan's Restaurant is a much better place than McDonald's. It is about 600 feet west.

ME 'n ED's

Adjacent to the airport is ME 'n ED's pizza parlor. They serve good pizza. Across the highway is McDonald's and Ryan's. Don't forget to try Country Waffles next to the Shell station.

Madera (Madera Muni. - MAE)

Madera Muni. Golf Course Snack Bar

Sadly, the Madera Airport Cafe is gone. There is an alternative. The Madera Municipal Golf Course is about two miles from the airport. You'll see it from the pattern just beyond and to the left of runway 30.

To get there, you can borrow a car from one of the FBOs and take Aviation Drive out to Avenue 17 and turn left. Go for about a mile and a half to the corner of Ave 17 and Rd 23 and you're there.

The snack bar is in the back side of the country club, with picture windows commanding a view of the driving range and the starting tees. There are about 15 tables or so and a TV that is permanently set to the Golfing Channel above the bar. The snack bar is more "bar" than "snack"; after all, this is a golf course.

The grill serves burgers or sandwiches with french fries and a soft drink or iced tea for about $5.00. Service was OK, but you may have to wait a bit if the waitress is delivering a cell-phone order to golfers on the 4th hole. For those who miss the grill hours (it closes at 3:00 PM) there is a fridge with premade cold-cut sandwiches and bags of chips, but they are not nearly as satisfying.

The food was good, served hot and quickly. It's a nice relaxed atmosphere, and the prices were reasonable. The only negative is that it isn't at the airport.

Mammoth Lakes (Mammoth - June Lakes - MMH)

Airport Cafe

If Chuck is in a good mood and not out driving his taxi, he will whip up a fine deli sandwich for you. If you ask nicely, he will brew a fresh cup of Italian/French roast coffee that is wonderful. He also will talk your ear off. If he's not in a good mood, the sandwich is still good, the coffee not fresh, and the talk nonexistent. The FBO will make a fresh cup of coffee if Chuck is cranky!!

Merced (MacReady Field - MCE)

Bobbi's Fly-In Cafe 

Merced, California, has a restaurant called the Fly-In Cafe. They are only open for breakfast. The food is very good, you get a lot of it and it is very inexpensive.

Monterey (Monterey - MRY)

The Golden Tee 

The Golden Tee at the Monterey Airport ILS has been there for years and was redecorated in 1996. It is located on the second floor of the terminal building itself and has a great view of the Monterey Bay and of takeoffs and landings if they're using runway 10, and of the taxiing airplanes if they're using runway 28.

They serve some of the best sand dabs I've ever tasted, lightly breaded and very delicate. Their Monterey Combo offers both the sand dabs and a calimari steak. The chef salad is very good, as is the BLT, which I always order with avocado. Don't miss the house salad dressing and the freshly baked bread. They're open for lunch and dinner. I don't know about breakfast.

Tarpy's Road House 

For a nice meal, consider Tarpy's in Monterey. A five-minute cab ride from Del Monte East is the FBO's light-plane tie-down area. Ask them on Unicom to call a cab, and it'll come right in and meet you just about when you're done with your plane.

The Del Monte Express 

I agree that Tarpy's Road House is excellent. However, if it's really a burger you want, take a cab the same distance in the other direction to Fremont Street and The Del Monte Express, "home of the five napkin adult hamburger." Long-time favorite of locals.

Napa (Napa County - APC)

Jonesy's Famous Steak House 

The annual Best of the Napa Valley voting picks Jonesy's at the Napa County Airport year after year in the steak house category. They have coffee shop food also.

AMENDMENT

I have been flying in to Jonesy's Famous Steak House since 1960. They had a sirloin steak dinner for two for $11.75. It was a great bargain then, and with inflation covering 37 years, it is now $18.75. They take a huge sirloin and cook it only one way and cut it into two portions. It is served with special cheese-covered potato strips and a huge green salad.

We either fly in or stop by driving back from Reno. It has the greatest food, bar none, and of course, being Napa, a great wine list.

Oceano (Oceano - L52)

Fish and Chips

Oceano CO (L52) is located between Santa Maria, CA, and San Luis Obispo, CA, and sits right near the dunes on the beach in Arroyo Grande. The strip is all of 2,300 feet paved and can get windy at times from what I hear. There are two+ great restaurants within two-three blocks, one of which is a great little seafood place. The FBO operators are very friendly and will be more than happy to give you directions to the restaurants.

Oceanside (Oceanside Muni. - L32)

Burger Bar

Oceanside Airport, south of John Wayne in Orange County, is a good place for a unique fly-in. The burgers are a three-block walk, across a busy highway, at a bar up the street perpendicular to the airport. They're friendly and will set you up with a good, greasy burger at almost any time of the day. Pay close attention to airspace and contact UNICOM early for landing advisories, as they're trying desperately to keep the airport open.

Ocotillo Wells (Ocotillo - L90)

Burro Bend Cafe

Ocotillo Wells airport is located 23 miles DME from the Julian VOR on the 074 radial. It is 12 nm east of Borrego Springs. The field is graded on a dry lake bed with two intersecting runways, 13-31 and 6-27. The few times I have been at the airport the surface has been excellent, although I still recommend a low pass to evaluate the surface.

The field is surrounded by desert. It is a great place to fly low; just be alert for ultralights, motorcycles and dune buggies. Turbulence picks up in the afternoon so I prefer to fly in for breakfast and be back over the mountains before it gets too bumpy.

After you land, taxi to the south side of the lake bed and look for the restaurant. There are tie-downs available. The restaurant is called the Burro Bend Cafe. They are open from 7:00 AM to 7:00 PM. You can call to check the runway conditions (619-767-5970). They aren't pilots, but they will tell you if planes have been flying in or if it looks wet.

The restaurant is full of out-of-the-way charm. Don't forget to ask for the pilot's guest book so you can tell everyone you were there and check on your friends.

This is my favorite $100+ hamburger place. I think it will be yours too.

Orange County/Santa Ana (John Wayne Airport - SNA)

Chanteclair's

John Wayne Airport is the only commercial airport for Orange County, CA. It is one of the 10 busiest in the country. There are left and right runways which effectively separates the small fry from the big iron. If you're flying a 737, taxi up to one of the terminals, walk down the jetway and into the terminal area. There are several popular fast-food places in the terminal, with prices only 130% of what they usually are.

Otherwise, you will have to land at one of the FBO's GA transient tie-down spots and avail yourself of the nutritious fare found in several vending machines placed strategically around the area. Bring exact change.

For real food, you will have to depart John Wayne Airport and go across MacArthur Boulevard. It may take a taxi to do this. Many restaurants and hotels are within sight, offering a menu for most any palate or budget. If you're throwing caution and your pocketbook to the winds, try Chanteclair's on MacArthur Blvd. for excellent French cuisine. Too bad you'll have to skip the wine list.

Oroville (Oroville Muni. - OVE)

Table Mountain Golf Club 

The restaurant at the golf course adjoining the Oroville Airport is quite good and only about 150 feet from aircraft tie-downs. The decor is typical golf course, but they have sit-down service and the welcome was enthusiastic. The food was good and the portions large. Their hours are 0800 to 1500 Mon. to Fri. and 0700 to 1500 Sat. and Sun. The golf course is named Table Mountain but the airport is in the Sacramento Valley and anything but mountainous. The golf course and restaurant are on the opposite side of the airport, across from the FBO. You can taxi right to the golf club where you will find plenty of parking and tie-downs.

AMENDMENT

Dropped in to the golf course restaurant at Oroville (OVE) this morning for breakfast and found the hash browns particularly good. Full breakfast with coffee was less than five bucks. Aircraft parking is west of RWY 1-19 and less than 100 yards from the golf course.

Oxnard (OXR)

Airport Cafe 

We tried the restaurant at Oxnard Airport in Oxnard, CA (OXR). It is located on the field next to the main terminal. The new owners have changed it from a steak house to an Italian restaurant. The food, salmon with Fettucini Alfredo, was dry and tasteless. From 6:00 PM to 7:30 PM we were the only customers in the place. Tell you anything?

Taqueria El Tapatio 

We skipped the airport restaurant and tied down at the FBO on the far east end of the field. A five-minute walk took us to Taqueria El Tapatio, a Mexican fast food place. The food was fresh, the price was right and we got to hang out with the locals. They even have a small patio from which to observe airplanes on short final. I recommend it to anyone interested in a great lunch at a relatively uncrowded airport. For those not interested in Mexican, there are a variety of other restaurants within easy walking distance, including possibly the only McDonald's decorated with pictures of general aviation aircraft.

Palm Springs (PSP)

Airport Restaurant

This is a very nice airport with light traffic. The shorter of the two strips will land you nearest to the Millionaire FBO where there is a transient tie-down. From there, it's a short walk to the restaurant and many good sandwiches. There is a pool out back of the FBO. I have seen many a pilot take a plunge before departing, so bring your trunks! Remember, this is on the east side of the field, not the side with commercial aviation! Also a "new" aircraft museum just opened and I highly recommend it! It's also on the east side, just a short walk from the restaurant! The people are very friendly at both!

Palo Alto (Santa Clara Cnty. - PAO)

The Palo Alto Hofbrau and Catering Restaurant

The Palo Alto Hofbrau and Catering Restaurant is adjacent to the airport. It's an easy 10- minute walk from the transient parking area. It's cafeteria style with excellent food and reasonable prices, worth a stop for sure.

Scott's Seafood Grill

There is a good restaurant within a 15-minute walk from the Palo Alto Airport. Keep walking westbound on Embarcadero Road, past the traffic light, and you will find Ming's. Across the street is Scott's Seafood Grill. It's a longer walk and will cost a few dollars more compared to the Hofbrau Restaurant at the golf course, but it's worth the effort.

Paso Robles (Paso Robles - PRB)

Ameila's Restaurant

This is an easy flight from the Bay area. Head down the Salinas Valley, and then hang a right. Careful of the Army National Guard restricted area to the north as you follow 101. There is a restaurant on the field that has additional outdoor seating if the weather is nice.

If you've never flown into PRB, be aware that it's one of the busiest, if not the busiest, uncontrolled fields in the state. It's quite an experience, and there tend to be lots of NORDO aircraft in the pattern. Of course, if you did your flight training in this area, odds are at least pretty good that you got to PRB on some cross country or another.

Perris (Perris Valley - L65)

Airport Cafe

Perris Valley is another well-kept secret, if that's possible in Southern California. The airport is primarily used for skydiving and ultralights, but real pilots are welcome. The skydiving is world class and great to watch. This weekend we watched sky surfing, snowboarding with a parachute, being taped for viewing on ESPN and NBC sports. The kids enjoyed watching the chute packing and ground practice sessions, while Dad watched the jumperette action.

The restaurant and sports bar were completed just two years ago and are really attractive. Food is American and Mexican, with daily specials. The prices are very reasonable. There is also a large lawn and swimming pool area for your use. Both are free!

When you fly in, remain west of the field and call Unicom on 122.9 at least five miles out. The jump zone is east of the field and the ultralights remain generally south. The runway is in good condition and paved for at least 3,000 feet and then extends to 5,100 feet with smooth dirt. This is a great place for a spring or fall 1/2 day excursion.

AMENDMENT

Feels like you're in a Formica food factory from the moment you walk through the door. The atmosphere is strange, weird vibes from the regulars. "You fly in airplanes and you DON'T jump out of them?" We left feeling like we had visited another planet.

Petaluma (Petaluma Muni. - O69)

29er Diner 

The Two-Niner Diner at Petaluma, CA, is a nice spot for a lunch and some airplane watching.

AMENDMENTS

Airplane watching is tops. The ramp is always busy. Food is usually good, but the service of late has been spotty. I waited 40 minutes for a hamburger and it was prepared wrong, 20 more minutes to get it right! No kidding, my friend was finished eating by the time I was served. Not so much as an "I'm sorry" from them.

I find the information on the service at Petaluma Airport in California (O69) not to be valid. I'm based at Petaluma and have been for more than four years. The service and the food is usually, almost all the time, very good. It might have been an off day for them, for it surely isn't the norm. The owner and workers at the restaurant are polite, courteous and friendly. Please give them another try.

Quincy (Gansner Field - 2O1)

Morning Thunder Cafe 

The Airport Steakhouse that used to be on the field is closed permanently. There are no plans by the county to place another restaurant operator on the field, as three successive restaurants have failed.

For breakfast or lunch, try the Morning Thunder Cafe in Quincy. It's a short, pleasant walk from the transient ramp, about one-half mile. Walk towards town from the airport, turn left at the Unocal 76 station onto Lawrence Street. The cafe is on your left about 100 yards ahead. It is open from 0700 to 1400.

AMENDMENTS

I would have to second the vote for the Morning Thunder Restaurant, the absolute best place for breakfast I've found in northern California. The flight out, westbound, down the Feather River Canyon, can be simply beautiful in the winter!

✈✈✈✈✈✈✈

Morning Thunder truly deserves to be bumped up in the ratings. I would give it a five for food and prices. Service is good too. It is a half mile walk from the airport, as previously noted, but worth every step.

Red Bluff (Red Bluff Muni. - RBL)

BJ's

Late last summer, I flew my Cessna 140 from Houston to Seattle to start a new job. My favorite restaurant along the way was BJ's Airport Restaurant in Red Bluff, CA. Aside from offering basic western American food, they have terrific service. Located on the second floor of the small terminal building at RBL, the wrap-around windows allow great airport views. I especially recommend their omelets. They are open EARLY for breakfast.

AMENDMENT

I flew to Red Bluff this morning with my six-year-old son for breakfast. The service was good and the food was excellent. I had the ham and eggs. My meal was very large and the price was very low, compared to any other place I've been. My son had the Mickey Mouse pancake. The total price was $6.72 for both of us. I was extremely satisfied!

Redding (Redding Muni. - RDD)

Peter Chu's

I was at Redding recently and discovered one of the first airport restaurants that I've seen offering only Chinese cuisine. It's upstairs from the passenger terminal, has huge windows so you can watch air operations and a very nice atmosphere.

Prices are very reasonable and they give you a lot of good-tasting food for the money. The staff are quick and courteous. Try it out, well worth the effort to go, especially if you like Chinese food!

AMENDMENT

On our way back to STS from Oregon, we stopped in Redding and checked out Peter Chu's. The food is good, a nice change of pace from burgers and other fried food you always find at airport diners.

Redding can get really hot, and if you stop for lunch and mention you're a pilot, they'll let you hang out in the air conditioning. Great views from upstairs; there's even an elevator. When you contact ground, tell them you want restaurant parking, if available. It's much closer than regular transient.

Airport Deli 

We were a little early on a flight from Oregon so we tried the deli in the IASCO Flight Center instead of the Chinese restaurant. The menu is limited to sandwiches, but the price is good and the help very friendly. Within the building you have access to a computerized weather service, FSS telephone and a counter to order your fuel.

Rialto (Rialto Muni - L67)

Dee-Dee's Airport Cafe

They have the best breakfast and lunch menu that I have found. The ham steak and eggs breakfast for $4.95 is especially good, a must try! You can park your plane right at the front window and admire it while you eat.

AMENDMENT

The worst meal we ever had was at this place. Dirty dishes, multiple! BAD service, wrong orders, and mediocre food. I know everybody can have a bad day, but we'll stay home if this is the only place open.

Riverside (Fla-Bob - RIR)

Airport Cafe 

Last of the great old-time experimenters hangouts and one of the most colorful cafes on any airport anywhere.

AMENDMENTS

A young couple just bought the cafe here and are spending a bundle on restoration. The place was an old WWII building moved here from March AFB and has lots of charm, including wood-beamed ceiling, river rock fireplace and great food. This is one of Southern California's best-kept secrets.

✈✈✈✈✈✈✈

It's like the airport that time forgot. The cafe has some great Southern cooking!

Riverside (Riverside Muni.- RAL)

Airport Cafe 

There is a fairly new restaurant in the terminal opened by the folks who ran the Fla-Bob eatery for years. This is the current favorite of my bunch, regular fare and good service.

Rosamond (Sky Park - L00)

Runway Cafe 

There is a small airport about 10 miles south of Mojave, California, with a great Mexican-style restaurant. The airport is called Rosamond Sky Park and is a combination fly-in community and public airport. The restaurant is right on the taxiway when you turn off midfield. You can taxi straight into it. The food is excellent. We went on a Sunday and it turns out they have an all-you-can-eat buffet on weekends for $8.95. This includes Belgian waffles made to order, omelets and a wide range of Mexican foods. There were also some very good desserts, like apple pie and chocolate cake. Enough variety for anyone, I would imagine.

I was really impressed with the food. My friends and I go somewhere every weekend and all agreed that this was one place to put on our list of places to visit frequently.

AMENDMENT

I can second the phenomenal Sunday brunch. Make sure you fast on Saturday and leave the baggage at home. You will eat plenty and need all the weight margin you can get on the way out. The food is truly excellent, service is great; there's a large patio right on the runway, and you park immediately in front of the restaurant.

Sacramento (Sacramento Executive Airport - SAC)

The Tailspin

Run by Cindy Gray and Marge Papp, the Tailspin Sandwiches Deli is a small, very friendly deli in the main terminal building at Sacramento Executive Airport. It is open from 8:00 AM to 3:00 PM, Monday through Friday, and from 9:00 AM to 3:00 PM on Saturday in the summer (March through October).

For breakfast, try their Sunrise Sandwich, egg, bacon and cheese on an English muffin, or a breakfast burrito, egg, refried beans, cheese and salsa, or one of a selection of doughnuts, muffins or bagels.

The lunch menu includes burgers and specialty sandwiches. Side orders include chili, soup, nachos and salads. Beverages include coffee, tea, iced tea, milk, lemonade, soft drinks and fruit drinks from Ocean Spray.

Friendly service makes this a nice stop for a bite to eat.

AMENDMENTS

After picking up your lunch at the deli, you can now sit in the bar and dance area of the Red Baron Restaurant upstairs and enjoy the beautiful view of the ramp and runways. It brings back memories of the days when the restaurant was in full swing. Very nice!

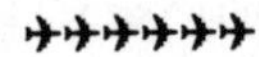

They have new hours: Tuesday - Friday 8:30 AM to 3:00 PM, Saturday - 9:00 AM to 3:00 PM, Sunday and Monday – Closed.

The Stick & Rudder 

The Stick & Rudder Restaurant officially opened for lunch and dinner 19 June 97. It is in the terminal building at the same location as the old Red Baron.

Salinas (Salinas Muni. - SNS)

Salinas Airport Bar & Grill

I was feeling like a long cross country — hadn't done one since I was a student pilot, so I flew from Beale AFB, CA, to Fresno and Salinas in a T-41C, which is a 210 HP C-172! The restaurant is located right on the field in the terminal building. It specializes in Asian cooking with some seafood. I had the calimari steak. The place has a bar and from what I saw was frequented by lots of local nonflyers.

San Carlos (San Carlos - SQL)

Sky Kitchen

This is an unpretentious diner for breakfast and lunch. It has a good pancake sandwich and nice potatoes.

AMENDMENT

In additon to the Sky Kitchen, San Carlos Airport now has a Burger King. The BK was built next to the Hiller Museum.

San Diego (Gillespie Field - SEE)

Sky Pup

There is a new restaurant at Gillespie Field, San Diego, the Sky Pup. It opened 5 April 1996. Their specialty is baking, but a regular menu is also offered. I had the burger with avocado and other trimmings. It was excellent! Open daily from 0800 to 1600.

San Diego (Montgomery Field - MYF)

Casa Machado

In the San Diego area there is a Mexican restaurant overlooking the runways at Montgomery Field (MYF). It is worth a stop.

94th Aero Squadron

The 94th Aero Squadron is located on the north side of the airport, near the control tower and FSS. Aircraft parking is available next to the fence. Ask ground control.

B&B Deli

B&B Deli is located across the street from Gibb's Flying Service. The menu has a wide selection of breakfast and lunch items.

Four Points Hotel 

There is no direct access from the airport. The best bet is to park at Gibb's Flying Service and call the hotel for a ride. Otherwise, the walk is less than a mile.

San Jose (San Jose Intl. - JSC)

94th Aero Squadron

Quite an interesting place with lots of old airplane photos. Headphone jacks are available so you can listen in on the tower frequency. They've got some VERY good sandwiches, priced moderately for an airport.

AMENDMENTS

I live here and LOVE the place. It's a tad pricey, but the overpricing amounts to less than 0.1 hours on the tach, so who cares! Certainly, try to make it out for Sunday brunch. The buffet is outstanding, with shrimp, marinated mushrooms, an assortment of fruit juices and an omelet bar. Tell the chef what you want; he makes it.

The table-mounted headsets let you listen to a combined clearance, ground, tower and approach/departure, so it's never dead. Kids and grownups love it.

You can count on a wait on days like Mother's Day, Father's Day, Easter, etc.

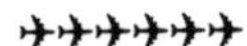

I flew in to have lunch at the 94th Aero Squadron for the first time 4/4/97 and was very disappointed. The restaurant is not easily accessible from the airfield. You have to exit and walk down the street to get to the front door.

Since we were unfamiliar with the airport, we were directed to park at the San Jose Jet Center several hundred yards further. We later discovered that we could have parked off intersection C and taxiway V.

We found the food overpriced. I guess you have to pay for the view of the airfield and the bells and whistles. Bottom line: It is not user-friendly for pilots.

San Luis Obispo (McChensey Field – SBP)

The Spirit of San Luis 

This right-on-the-airport restaurant affords an excellent view of bounce and goes. Sunday brunch is great with champagne for passengers! Try the Mile High Club for a good sandwich.

AMENDMENT

A great place for business lunch or Sunday brunch. I have been there many times, always good food. Try the bread bowl clam chowder.

Pull right up to the restaurant. It is close to the base of the tower.

Santa Barbara (Santa Barbara Municipal - SBA)

The Beachside Grill 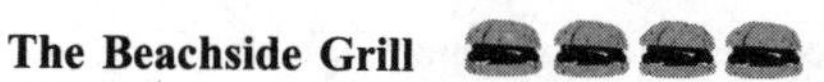

The Beachside Grill in Santa Barbara is special. Fly into SBA and park at Mercury FBO. It is about a 10-minute walk to the beach from there. The friendly personnel at the desk will give you easy directions. It's one of the few airports that I know of where you can actually walk to a restaurant off the field.

The Elephant Bar 

Santa Barbara Airport has another restaurant called The Elephant Bar. GREAT FOOD! The atmosphere is like an African safari. The prices are midrange. You can taxi right up to it. Just ask ground to point the way. If you are not flying out, try the large COCO LOCO and then take a cab to a motel.

AMENDMENTS

The Elephant Bar in Santa Barbara is famous for their mud pie. Try it, you'll love it. Check your weight and balance before you depart; however.

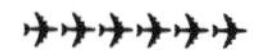

The Elephant Bar is still a great restaurant. The FBO adjacent to The Elephant Bar is now operated by Signature, with HIGH ramp fees or minimum fuel purchase, especially for twins: 60 dollars or 60 gallons for a Cessna 414. I thought the 100 dollars was supposed to go for avgas and tips.

✈✈✈✈✈✈✈

The Elephant Bar is still a fun place to go, but it has been spoiled for the flying crowd by the presence of Signature. Their exorbitant ramp fees and surly attitude are such that I cannot, in good conscience, recommend that you use their facilities. If you must come to The Elephant Bar, I suggest that you park at Mercury and take a taxi.

Silver Wings 

A very nice, inexpensive restaurant, with good food and an aeronautical theme is the Silver Wings in the main terminal. It is a very short walk from Mercury Air Center. The restaurant is upstairs, and the patio dining area offers a great view of the runways and the terminal ramp, not to mention beautiful mountain vistas.

The restaurant has a number of classic photos from Santa Barbara's aviation history. Did you know that the first Lockheed plant was here before World War I, when the family used the original spelling of Loughead?

Cajun Kitchen 

I flew into SBA early on a Saturday morning only to find out that The Elephant Bar is closed until 1100! So we asked at Signature about other restaurants and were told about the Cajun Kitchen.

Cajun Kitchen is a short walk from Signature, out the back door, past The Elephant Bar, right onto the main road, right at the light. You'll find Cajun Kitchen in the Airport Plaza on the left side of the street.

A pleasantly busy little spot. The scrambled eggs with chilies were good, and the home fries were EXCELLENT!

If you're looking for an inexpensive breakfast at SBA, try the Cajun Kitchen, you won't be disappointed.

Santa Maria (Santa Maria Public - SMX)

Pepper Garcia's

There is a good, sit-down Mexican restaurant on the north side of the field directly on the ramp. You can park almost in front, and a quick walk around to the side brings you in the front door. SMX ground control is happy to provide help finding it.

AMENDMENT

They also have excellent hamburgers for any Gringos in the group. Pepper Garcia's is in the main passenger terminal at SMX, just a few steps away from the GA transient parking area. Open for breakfast, lunch, and dinner.

Summit Steak House

Also on the airport at SMX is the Santa Maria Airport Hilton. Inside is the Summit Steak House — dinners only last time I checked. There's another area where breakfast and lunch are served. You can park your airplane right at the back door of the Hilton.

Santa Monica (Santa Monica Municipal - SMO)

DC-3 Restaurant

The DC-3 is fancy to the point that reservations are recommended. The museum at this field is top drawer!

AMENDMENT

WOW! The DC3 and the Museum of Flying are just great! They have a tortilla soup on Thursday that is as good as I have ever found! General manager Dan Ryan and executive chef John Wedig, nominated for Chef of the Year, tour the tables to make your dining experience truly wonderful. The DC-3 just received its four-star rating, so you know the food is great. My fave is the cheeseburger and fries, fries better than Micky D's. If you get to SMO, try DC-3 and check out the museum — great fun and great food.

Typhoon

Another great place is the Typhoon in the administration building, directly across the field from the tower. This is a REAL restaurant, but not at all stuffy and not as expensive as DC-3 across the field. The food is a truly interesting Thai/Chinese/California mix. It has a great closeup view of the runway. There are parking spaces on the ramp directly in front, and the management always finds room for fly-in customers, even though it gets pretty busy with off-airport people at night.

I'm based at SMO, and this is everyone's regular spot to celebrate having walked away from another one. They offer fun, Trader Vic-type tropical drinks and a long list of Asian beers.

The Spitfire Grill 

The best is the Spitfire Grill, just a short walk (five min.) down the street from the transient parking/administration building. This is the local pilot's hangout. They have a somewhat standard, but creative menu, a nice patio, and it's definitely the most affordable of the three places.

Santa Paula (Santa Paula Airport - SZP)

Logsdon's

Here's an old airport with a single runway and no lights, although night landings are permitted. They have a relatively new restaurant. It is a popular hangout for the locals and has a full bar and a full menu — very nice atmosphere. Aviation art for sale adorns the walls. Some tables face the runway and aircraft parking. The prices are reasonable, the food is good, but the service is slow. It's a small town, so try not to be on a schedule. When you're done eating, walk around the airport. This is a haven for antiques, WACOs, PT-17s, Howards, Moths, etc. There is even a Spartan, the only one I have ever seen! There is an "open house" one weekend every month. The local owners pull out their classics for public view. These people are wonderful and will talk you hoarse. I believe it is the first weekend of the month, and Sunday is the most popular day.

A word of caution. I got my license at this airport and consider it my second home, but there has been a rash of accidents recently. This has been mainly due to neglect by visiting pilots, not the design or layout of the airport. When visiting, follow standard pattern procedure, and if you have a radio, use it. NO STRAIGHT-INS. The place is small but always busy on the weekends, and there is no tower.

AMENDMENT

Night landings are NOT allowed. The California DOT made SZP remove even the reflectors that lined the runway. There are people who do make night landings, but it is at their risk and they might have problems with their insurance.

Santa Rosa (Sonoma County - STS)

John Ash

John Ash in Santa Rosa is a cab ride from the airport. It's one of the best restaurants in the area, with excellent food, a nightly wild game special, interesting salads, and great desserts. They've an extensive assortment of good wines in their cellar, for the passengers, or for those staying overnight in the area.

Sellini Bar & Grille 

Very nice "California" fare, with a little Greek flavor. A little upscale, but worth it. The ILS makes it a lot easier to get into.

AMENDMENTS

I just flew up there last weekend and paid my first visit to the place after receiving numerous recommendations. The food is WONDERFUL. I had one of the specials that day: Cajun Prime Rib. The Cajun sauce they used was very good, not too hot, but definitely spicy. It was served with garlic stir-fried zucchini and wild rice.

My companion had the skewered chicken, which is basically a shish kabob made with chicken. It had a light cream/wine sauce that was really tasty.

Portion sizes are just right, and I'm a big person, so that probably makes them "hearty," and the price is quite good for the quality. We spent just a little over $30.00 for dinner for two. This included one glass of the house white wine, which I was told is very good.

Granted, it's not just burgers, but it's extremely good.

The restaurant is in the airline terminal at STS. If you park in GA transient, you get to go through a gate that locks from the outside. In order to get back to your plane, you'll want to make note of the sign on the back of the gate saying what the combination to the lock is. Note that as airline terminals go, this one is very small, five scheduled flights a day on United Express to SFO. I didn't chat with the gate person to figure out how long it really took, because she was busy with customers the first time through and closed after we finished dinner.

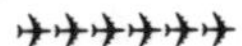

The Selini Grill on-field at Santa Rosa is my favorite stop. They have a great menu, including pasta and seafood. On the weekends you can sit outside under an umbrella table on the grass, listen to live jazz and watch the airplane action. The only thing better would be to be a passenger so I could partake of the Bass Ale on tap.

My favorite lodging in town is the Doubletree Inn on top of a hill overlooking the vineyards. Good restaurant, nice rooms, and beautiful pool area for $79.00 to $99.00.

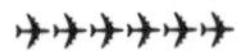

The Sonoma County Airport handles light jets and everything smaller. They also have Sheriff's Rescue Copter, Long Ranger II, and a Aerospatiale Hospital Copter. They have a new restaurant there called Selini's that is very good. It gets more business from the locals than from travelers.

Santa Ynez (Santa Yenz - IZA)

Solvang's Bakeries 

The scenery is spectacular and the quaint little landing strip is just downright fun. I don't remember there being a restaurant on the field itself, but they have taxis, dial-a-ride, and a bus that take you into Solvang, a Scandinavian township, within a couple of minutes. Spending the day is worth it, if only for the bakeries.

AMENDMENTS

Only junk vending machines at the airport. There's a cab, $10.00 for three people one way, very friendly mom-and-pop operation. Great food in town.

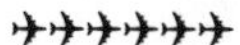

There's also a casino nearby on the local Indian reservation for those so inclined. Watch out for glider traffic! The gliders land in a dirt area in-line with but just short of the approach end of the western-facing runway 25, if I remember right.

Santa Ynez also has some very good restaurants that don't require a cab ride into Solvang. Just walk out to the main road past the firehouse and turn left, then the first right into the town. The main street is maybe a 15-minute walk from the airport, and there are several restaurants that always have something to satisfy me. Furthermore, there are some really cool old-time stores. You may be able to go antiquing in addition to getting that ostrich burger. Enjoy!

Schellville (Sonoma Valley - 0Q3)

Viansa's Tuscan Grill

If you're a wine snob, or just like to eat well, fly in to Schellville's Sonoma Valley Airport. There are two runways: 07-25 is 2,700 feet and is the main runway in use; 17-35 is 1,500 feet and is marked restricted. Check with the airport office at 707-938-5382 before you decide to use the shorter runway.

Park on the grassy area in front of the faux ATC tower and walk about one mile south on Route 121 to Viansa Winery. It's on the left just past Cline Vineyards, also open for tasting, but no first hand experience to report.

Viansa has an extensive Italian marketplace with all sorts of gourmet food to sample before you buy. There is also a deli indoors, and of course there is wine tasting. Now you are not going to be able to fly home, but that is another story for another day.

Outside the marketplace and tasting building, Viansa has a Tuscan Grill that has several different specialty sandwiches averaging $6.50 — very tasty grilled chicken, Italian sausage, pork tenderloin, etc. Enjoy some wine with your Tuscan Grill or deli lunch. You can sit at one of several tables on top of the Viansa hill overlooking the airport. Then when you are done, you can have some gelato or sorbetto afterwards.

The only problem with all this food is you have to walk back to the airport afterwards. Be careful of the traffic on Route 121. It is very busy, especially on weekends.

Stockton (Stockton Metropolitan - SCK)

Top Flight Cafe

There is a coffee shop/restaurant in the terminal building at Stockton. Not great, maybe two burgers, but it's there if you wind up in that area hungry. Seems as if lots of folks head over here to practice instrument approaches. Those can sure make you hungry! You can sit outside on a deck and watch the air traffic. Sometimes the airlines go over there. It's interesting to watch DC-10s do touch and goes.

AMENDMENT

The restaurant, Top Flight Cafe, must have improved since the last report. It's full service and has specials each day when it's open, which is Monday through Friday. I'd give it an above-average rating. With no regularly scheduled airlines, the airport is a pleasure. The tower seems happy to have someone call. The restaurant is located in the terminal building on the second floor. Transient parking is immediately adjacent to the terminal off runway 11R.

Tracy (Tracy Muni. - TCY)

Mexican Deli

There are two runways, 29/11 4,000 feet + and 25/07 3,500 feet+

About 1/4 mile from the airport is an excellent Mexican Deli and store. It's open from around 0600 to 1800 hours. They have a couple of benches on the side. During the week the lunch hour is packed. Saturday and Sunday, brush up on your Spanish, the locals are out. I've been there several times now and every time has been a good experience! I love it!

Taco Bell only wishes!!!

Last time I flew out there I talked to the Parks and Recreation manager. He mentioned they were putting in a little park, with benches and barbecues.

Trinity Center (Trinity Center - O86)

Trinity Center Inn

Trinity Center is a great sunny day. The airport is on the edge of Trinity Lake (technically, Claire Engle Lake, but don't call it that around the locals), in the foothills of the Trinity Alps. There is no fuel available, but Redding is only a few minutes away.

On the airport is the Trinity Center Inn, a motel and restaurant. It serves dinner most of the time and lunch in the spring and summer. It is at the northwest end of the runway.

Sasquatch

Up the street a couple of blocks is the Sasquatch, a bar and grill that is open for lunch and dinner on weekends. The food is great for mountain fair, and the location is awesome. Enjoy!

Yellowjacket

There is one more eatery worth mentioning in Trinity Center. It's a little burger joint with an outdoor patio. The name is the Yellowjacket, after the local bee tribe. Good burgers and nice people.

Ukiah (Ukiah Muni. - UKI)

The Beacon 

The Beacon Restaurant is located across the street and about 1/2 block north of the Ukiah Airport. It offers good light restaurant fair served by nice folks. I have never gotten a poor meal there. They make a really mean club sandwich.

The field is an Air Tanker base, so keep your eyes peeled for S2Fs and Cessna 337s during the fire season.

El Azteca

If you're into Mexican food, try El Azteca Restaurant. It is adjacent to the airport. From the airport entrance, turn left and walk about 10 minutes. It's on the left. The food is great and the portions are huge. Lunch runs around $5.95 for most of their entrees. They also serve gringo hamburgers.

Upland (Cable Airport - CCB)

Cable Cafe 

I'll have to vote for Cable Airport (CCB), owned by Walter Cable. It's about 40 miles east of Los Angeles. An uncontrolled field, they call themselves the largest privately owned airport open for public use. Very friendly people, and their restaurant, the Cable Cafe, has outside tables with a view of the runway.

AMENDMENTS

The Cable Cafe at Cable Airport, the "World's Largest Private Airport," is the best. Located in Upland, CA, northwest of Ontario.

I look forward to each visit and plan around breakfast. The biscuits and gravy are especially GREAT! They can be ordered with anything and enhance whatever you order. Hamburgers are available. The service is excellent even during the annual January Open House and Airshow.

I highly recommend giving Cable Airport a try. It's open to the public year round and has all services. The runway is smooth and the mountains to the north are majestic on a clear day.

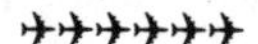

This is by far my favorite airport cafe. No matter the time, whenever I am there, I order the Polish sausage omelet with biscuits and gravy. The food and service are excellent.

Vacaville (Solano County - O45)

The Coffee Tree 

Even though they closed down the restaurant last year, with mixed rumors about reopening: (a) as more outlet malls (b) as the restaurant with more theme park (c) never.

The Nut Tree is still a fine fly-in for lunch or dinner.

Not only is it a lovely park, and so much more peaceful, now that it's closed, but it's a lovely airport with an instrument approach.

Just a short walk over I-80, there's a foot bridge which takes you to the Coffee Tree. While it's not the great food they had at the Nut Tree, it's still pretty decent fare at a reasonable price. If you like omelets stuffed to overflowing, as I do, this is your place.

So, don't bail on The Tree! It's still there, sorta'.

If you feel up to a slightly longer walk, we're talking nearly a mile, maybe 20-30 minutes worth, there are a few other restaurants up the frontage road.

My wife and I ate the very last two meals served at The Nut Tree. We went there on a whim, inspired by "aren't they supposed to closedown in 2 or 3 weeks?" It turned out that they were closing that night. We got there late, right at closing time, but everyone was so happy to have another customer that they seated and served us. It was delicious, but a little spooky to see the stores all cleared out like that.

Watsonville (Watsonville - WVI)

Zuniga's

Located on the field. They serve better-than-average Mexican fare. In fact, it's good!

AMENDMENTS

I have just a little more to say about the place on the field in Watsonville. Zuniga's is one of the best places in the Monterey Bay area for Mexican food. Great salsa, one of my personal criteria for judging good Mexican food. The portions are more than generous, especially if you get the super burrito. When they say super, they mean Super. It's huge!

Not only is the food at Zuniga's good, but the restaurant itself is one of the most convenient fly-ins in the southern Bay Area, with tie-downs immediately in front of the restaurant. Zuniga's is open every day until 9:00 PM.

Zuniga's is closed. I flew there for lunch last Monday (10 Feb. 97). A lady at the airport informed us that they may reopen in a month or so. You can confirm this with a telephone call.

Zuniga's is open again!!!! Enjoy.

Willits (Ells Field - O28)

Brown Bag It

Like trains? Fly into Ells-Willits (O28) and ride the famous Skunk Train to the Mendocino coast. This is not a trip for those in a hurry. A round-trip ticket, an all-day ride, is only $28.00. The folks at the airport will give you a ride to the station if you ask them nicely. Pack a lunch!

Willows (Willows - WLW)

Nancy's Airport Cafe

Well it's a little outside the Bay Area, but there is a cafe at the Willows Airport I once flew to for breakfast. Their breakfast was great and the price very reasonable.

AMENDMENTS

Nancy's Airport restaurant is open 24 hours a day. The food is very good and the prices are very reasonable. My favorite is the French dip sandwich, which includes your choice of potatoes.

Do not expect any answer on the Unicom, as this is a county-run operation and they do not give advisories.

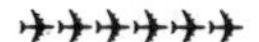

The food at Nancy's is very good, reasonably priced, and the service is good considering the number of LOCALS who frequent the place. If your significant other is a little uppity, maybe

you should go someplace else. The locals definitely consider it their territory. This makes a good stopping place en route to Oregon or just a low-key meal from the Bay Area.

Woodlake (Woodlake - O42)

The Outpost

Woodlake is a small airport nestled at the foot of the Sierras about 15 minutes north of Porterville. The restaurant on-field is the Outpost, and it looks like one. It resembles an old wooden cabin. The food is good, with a wide selection. Pies are the main dessert and are baked fresh daily on site. The waitresses are friendly and funny. Expect to be kidded a bit if you give them an opening. We've made several trips there and haven't had a bad meal. The pies are delicious.

Last time we were there, the waitress said the airport was for sale. Anybody wanna' buy an airport?

Woodland (Watts-Woodland Airport – 041)

Flier's Club 

Watts-Woodland Airport has an excellent little restaurant called the Flier's Club. It is about 200 yards off the end of runway 36. I find it a little pricey but sandwiches are excellent and they offer light lunches.

AMENDMENT

The Flier's Club restaurant is off the north end of the runway and is part of the golf (private) course. It is a very, very nice restaurant, and yes, for a private club, the prices are a little high but well worth it. I'd give it five hamburgers.

Yuba City (Sutter - O52)

Cafe Molly D's 

There is no restaurant on the field. Walk out the gate on the north side of the field, turn left on the street, then make another left for about 1/4 mile and you will find Cafe Molly D's, a full service restaurant. They serve great breakfasts all day – four-egg omelets for under $5.00 and other goodies for those who like to eat a lot. I'll be back!

Aspen (Sardy Field - ASE)

Planet Hollywood

Flying into Aspen is an experience you are not likely to forget. First of all, the approach is a thrill a minute. When you walk around the ramp, you will think you have stumbled into a fly-in for jets. The flight in and out is over some of the most spectacular country you will ever see. BE CAREFUL, these are real mountains and you will have to be able to cruise well above 12,000 feet, so check density altitude carefully. Don't forget to take survival gear just in case. A walk out after an engine failure could take weeks.

I have never seen so many Lears, Falcons, Starships, etc., etc., etc., in one place in my life. The FBO is like a resort. The people are very friendly, and they will gladly loan you a crew car if one is left. Be nice and fill it as well as your airplane.

The town is about five minutes away and is crowded with great places to eat. The two most recognizable restaurant names in Aspen are Planet Hollywood and Hard Rock Cafe, but there are dozens of local joints that have more character. Generally hamburgers, deli sandwiches, chicken breast sandwiches, chili or Mexican are all plentiful. I'm sure there are more elegant places with higher prices, but I avoid such places like the plague. Parking can be a problem, but on Sunday the meters are free, so just park where you can. No place in town is really too far to walk. We were in a C-182 out of Jeffco (BJC), just across the Corona pass and up the valley to Aspen.

Buena Vista (Buena Vista Muni. - 7V1)

Cafe Del Sol 

The Buena Vista, CO, airport has a courtesy car so that you can traverse the two miles to town easily. Many restaurants are there, but the Cafe Del Sol is my favorite. It has great Mexican food, in the medium price range, and is right on the main drag. Next door is a wonderful bed and breakfast place. The B&B is expensive!

Buffalo Bar and Grill

Buena Vista Airport (7V1) is located in the Rocky Mountains of Colorado. Elevation is 7,946 feet with an 8,300-foot paved runway. You can choose between large heated hangars or several outside tie-downs.

There are numerous restaurants in town which are within walking distance. The airport provides three courtesy cars. Casa Del Sol does have great, moderately priced Mexican food. My favorite is the Buffalo Bar and Grill. Any red meat lover will feel at home at this moderately priced steakhouse.

Meister House

There are also several bed and breakfasts. The Meister House is a standout. This recently restored, historic hotel has six unique rooms. A truly gourmet breakfast is included.

The area is loaded with activities in the summer or winter, including skiing, hiking, camping, fishing and rafting on the Arkansas River. Mt. Princeton Hot Spring is nearby, with hot spring pools along the Chalk Creek.

The local FBO is staffed by terrific people who will do anything they possibly can to make your visit a memorable one.

Denver (Centennial Airport - APA)

The Perfect Landing

The Perfect Landing restaurant is on the field at Centennial Airport (APA) on the south end of Denver. The restaurant is upstairs in the Jet Center FBO building. They serve breakfast and lunch. The food is very good. The icing on this cake is the view! All the windows face west with an outstanding view of the runways and the Rocky Mountains. It is hard to beat.

The Final Approach

Just inside the lobby of the Holiday Inn at Centennial Airport you'll find the Final Approach. It's a bit of a hike if you don't park at the Denver Jet Center, but it's worth it.

The restaurant is straight Holiday Inn with a bit of an aviation motif, photographs and paintings of local warbird drivers. Happy hour specials are frequent and the lounge boasts a large selection of microbrews. This place is NEVER crowded, which is puzzling. The food is reasonable and terrific.

If you're into football, this is where the Denver Broncos, their coaches and wives hang out. It's a very quiet place to spend the night after a long cross-country!

Denver (Jeffco Airport - BJC)

Denver Air

The fly-in breakfast at Denver Air (BJC) every Saturday and Sunday morning is a great deal. While not really a restaurant, you'll enjoy two eggs, bacon, sausage or (both), pancakes, toast and all the coffee or juice you can handle for only two bucks!

Service is semi-self serve. The clubhouse atmosphere offers a lot of great hangar flying.

Erie (Tri-County Airpark - 48V)

Airport Restaurant 

Try Tri-County Airpark (48V) on the weekends for a great burger. The place seems to fill up around noon with local pilots.

Fort Collins (Downtown Fort Collins Airport - 3V5)

Charco Broiler 

The best hamburger and steak sandwich in Colorado is at the Charco Broiler just three blocks from the Downtown Fort Collins Airport (3V5). If walking isn't your style, ask the manager, Sharon, for her courtesy car and you are on your way. Try the New Yorker and ask for extra cream cheese!

Granby (Granby - Grand County - GNB)

Bertie's Restaurant 

A telephone call to 970-887-3661, alternative 970-537-8086, will get you the combination to a lockbox which holds the keys to the courtesy car. Be sure to tell 'em you're a pilot and ask for directions to Bertie's Restaurant (970-887-3632). It is about 1.5 miles away in Granby. There is a woodstove in the dining area and they serve really great STEAKS!

Greeley (Greeley Muni. - GXY)

Barnstormer Restaurant 

Try the Barnstormer Restaurant in the terminal building at GXY, Greeley, CO. Great burgers, breakfasts, and specials. It's open daily from 7:00 AM to 3:00 PM, although I heard that they close earlier on Sundays.

Kremmling (McElroy Airfield - 20V)

The Wild Rose Restaurant 

McElroy Airfield sits between two really nice restaurants. Each will come and get you for the price of a phone call. I don't have a favorite. They're both pretty good. A courtesy car is available. Dial 719-724-9407 for the free shuttle to The Wild Rose.

The Wagon Restaurant 

McElroy Airfield sits between two really nice restaurants. Each will come and get you for the price of a phone call. I don't have a favorite. They're both pretty good. A courtesy car is available. Dial 719-724-9219 for the free shuttle to The Wagon Restaurant.

Limon (Limon Municipal : LIC)

Flying J

The Flying J is a truck stop cafe. It serves good ol' American food at low prices. It is an easy walk from the airport, less than a half mile.

Pueblo (Pueblo Memorial - PUB)

Airport Restaurant

Pueblo Memorial (PUB) is a stable place that you can count on not closing, since it serves an air carrier. It's a notch lower in quality than Meadow Lake was. The atmosphere is more like a diner with good basic food. Very convenient to get to, with lots of service from the FBO.

Burlington (Johnnycake - 22B)

Gail's Country Kitchen

The restaurant is pleasant and offers diner-type food for a reasonable price. It is located right on the airport. The last time I was up in Connecticut, at Mountain Meadow Airport, aka Johnnycake Airport, was about eight months ago. I used to keep my aircraft there before moving to Florida.

This is a fun, general aviation field that understands taildraggers.

AMENDMENTS

Another fly-in restaurant bites the dust. Gail's Country Kitchen at 22B closed Sunday 7/7/96. No news on what is going in the facility. Gail is moving two miles away, but definitely not a convenient place to get into by air.

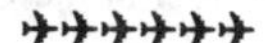

The restaurant on Mountain Meadow, aka Johnnycake Airport, has reopened under new management. The young man in charge is very enthusiastic and really interested in pulling in fly-in traffic to rediscover this place. As an interesting sidelight, they have commissioned a spectacular aviation mural on all the walls of the dining room. It depicts the local scenery and classic aircraft in flight.

The kitchen is getting its act together. The food we sampled, a grilled chicken sandwich and a $50.00 hamburger, were quite good. Why not a $100 hamburger? Well, we were only flying in from Chester, CT, which is a pretty short ride in a fast Bonanza. The food, atmosphere and accommodating service should help this place make a go of it.

Danbury (Danbury Muni. - DXR)

Hangar One Cafe

I go into Danbury fairly frequently. The Hangar One is no epicurean delight, but it is a good place for breakfast, a burger or a fast cup of coffee. I am glad it is there and hope it stays.

AMENDMENT

I stopped at Danbury Airport (DXR). The attendant at the airport, when asked for a place to grab something to eat, suggested a small shopping mall about 1/2 mile from the airport. Apparently the restaurant on the field is closed.

The Classic Rock Brew Pub

The Classic Rock Brew Pub has recently opened in the same location as the old restaurant on the field at Danbury (DXR). They're located on the north side of the approach end of RWY 26 (near the Reliant). Local phone is 203-792-4430 for more information or to check on their hours.

East Haddam (Goodspeed Airport and Seaplane Base - 42B)

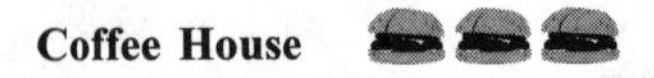

This is a short paved strip right next to the river. There is a dock for tying up floatplanes and a concrete ramp for amphibs. A short walk up the street leads to a charming coffeehouse and cafe where you can get sandwiches, chili, soup, and fancy soft drinks. I recommend it highly!

The Gelston House

In addition to the coffee house there is a great, but expensive restaurant called the Gelston House. We have eaten there many times, breakfast, lunch and dinner. It has always been wonderful, no hamburgers though. You can see a show next door at the Goodspeed Opera House. This historic site puts on several musicals a season. Both are in easy walking distance of the airport.

Groton/New London (Groton/New London - GON)

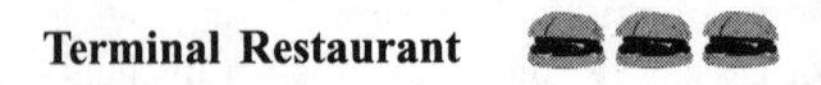

Always go to the Groton-New London restaurant at the terminal. It is nice all around. They seem to be open when all others seem to be closed.

Hartford (Hartford/Brainard - HFD)

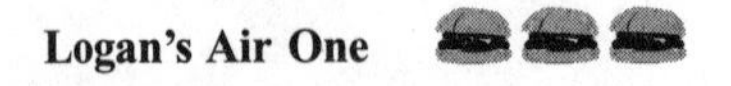

Logan's Air One is located in the same building as the MillionAire North FBO. Aircraft parking is right in front. They feature a variety of burgers, large salads as well as full meals. The seating overlooks the runway. In the summer a nice outdoor area is opened. It features round picnic tables with umbrellas.

Willimantic (Windham - IJD)

Greek Restaurant 

Walk up the hill behind and to the right of the main building to find a Greek restaurant offering all types of foods, healthy portions, and it is inexpensive. It always seems to be open, even on Mondays, when everything in New England is closed.

Bagel One 

Just a quick walk from Windham, Willimantic is Bagel One, featuring NY-style bagels, bagel sandwiches, soups, stews and a variety of coffees at reasonable prices! Open seven days a week.

Middletown (Summit Airpark - N92)

Howard's Family Restaurant

Howard's Family Restaurant (302-378-4429) is nearby with courtesy car service. Ask at the FBO desk for transportation.

Jean's Luncheonette 

This is your basic sandwich shop. It is very close to the airport. The FBO can arrange courtesy car transportation.

Wilmington (New Castle County - ILG)

The Dutch Pantry 

Wilmington actually has three restaurants on the airport property. I can't remember the name of all of them. The Dutch Pantry is my favorite. It is a great place for lunch. It has large windows, which afford a good view of runway 9-27. They also have wonderful apple pie! (302-322-4467)

The Air Transport Command

The Air Transport Command is located on the edge of New Castle County Airport, and its motif is WWII / 1940s. The entrance to the restaurant is lined with WWII hardware, jeeps and the like. The sound of planes taking off and landing nearby adds to the ambiance! Most tables give a view of the runways.

You can either leave your plane at the terminal and walk to the restaurant, about 1/2 mile along a busy road, or park the plane right by the restaurant. To park alongside the restaurant, however, requires special approval from airport security. The Airport Safety Department can be reached at 302-328-4632. To obtain it, you must either call ahead or stop at the terminal first. Approval cannot be given over the radio.

The restaurant serves a full menu, with entrees starting at about $12.00. Their specialty is beef, but they also have a nice selection of veal and seafood. Their warm, home-baked cracked-wheat bread was delicious! Dinner for two will probably run about $35-$40. (302-328-3527)

There are a variety of other restaurants bordering the airport, including the previously mentioned Dutch Pantry, as well as a TGI Fridays and a Wendy's.

AMENDMENT

Another plug for the Air Transport Command: excellent food, good prices, and "beer cheese" soup!

Jake's 

Please allow me to amend the ILG listing to include the best burgers in the greater Wilmington/ Philadelphia area. The treasure can be found at Jake's, a perennial winner of the Wilmington's Best popular vote conducted by the Wilmington News Journal. The good news is that Jake's went multi-site recently, and the new location is just west of the ILG terminal on the south side of Rt. 13, the airport frontage highway, approximately 1/4 mile beyond the intersection of the airport boundary road. If you fly the ILS to RWY 1, Jake's will be off your left wing as you cross the highway on the approach end of the runway. It's a tough walk, 1/2 to 1 mile, an easy cab, or try to con the FBO into a courtesy van ride.

Arner's Restaurant 

We parked by the terminal (there is no fee) and walked along the highway north for about 0.4 miles.

Most any entree is good. They are all reasonably priced. Veal parmesan with spaghetti is my favorite, though the crab cakes are a very close second.

Arner's is a very small chain of restaurants with only five locations, four of which are in Reading, PA, where the original restaurant was located. It has retained its family-operated style. Try the strawberry pie. If you go away hungry, it's your own fault.

Arcadia (Arcadia Muni. - X06)

The Deli

Arcadia is about 20 miles west of Sebring in the middle of the state. It is a $3.00 – $4.00 cab ride into town. The older part of town includes numerous antique stores. One of the more popular spots is the old Opera House. This upstairs facility is packed with antiques, including many of the old projectors, silent and early sound which have been used in the theater. The Deli up the street has excellent, large sandwiches and delicious fresh pies, $1.50 a slice.

Bartow (Bartow Muni. - BOW)

Airport Cafe

The restaurant is off-field about 1/2 mile. It is not open on Sunday. I left hungry and went to Lakeland to eat.

Bunnell (Flagler County - X47)

Wings Restaurant

Wings Restaurant at Flagler County Airport (X47) in Bunnell is open from breakfast through dinner. Good home cooking and nice people. It's not fancy but a real nice place for a fly-in meal. They have a German cook who makes a fantastic Wienerschnitzel as well as other authentic dishes, great for dinner. Drinks are available for the passengers and prices are quite reasonable.

Flagler Co. Airport is another good reason to visit. It has a real cross section of general aviation, with skydiving, gyrocopters, aerobatics and flight instruction, all in a friendly atmosphere.

Cedar Key (George Lewis - CDK)

Captain's Table 

Your basic tourist trap. Fly over the town to flag down the taxi lady. It was $5.00 or so into town. There are lots of places to eat or sleep over. This is the most amusing airport to watch the weekend warriors test their short field landings. 2,200-feet is that short?

AMENDMENT

My wife and I did an overnight trip to Cedar Key and enjoyed it. *The* cab driver, Lester Ridgeway, monitors CTAF in his cab and at home and will suggest a runway. His fee for the five-minute trip to town in the Checker cab is "whatever you think is right." If he is busy, his wife may come get you. The short field, with an over-some-low-trees approach onto a displaced threshold with a Gulf of Mexico overrun, is fun. Three cars came out from town to watch me land and drove off while I taxied in. There is no fuel or any service available.

Four or five seafood restaurants are in town with similar pricing, and some serve breakfast, lunch and dinner. We ate at the Captain's Table for about $50.00 including a drink each, tax and tip. The seafood was excellent. I would suggest calling ahead for a reservation and attempting to get a window seat for sunset. The restaurant is located on the second floor above a lounge.

There are several hotels on the island. Most are on the dock. If you want to, you can rent a golf cart to ride around, rent a boat to go out to the outer islands, or take an airboat tour of the area. The big events are an art festival and a seafood festival. They occur in April and October. There is no real beach on this island, although it looked like there were some nice ones on the outer islands. Have fun!

Clewiston (Airglades Airport - 2IS)

D.Z.'s Food Shack 

Fly into Airglades most any Saturday or Sunday and get a made-to-order burger. Air Adventures, the local FBO and skydiving business, has an individual come in and cook for the parachutists. Very, very informal, jeans and t-shirts. All service is at an outside table and stools. This is not a sit-down type of restaurant, but the food is good. The pattern is to the south side of field. When you arrive, ask for D.Z.'s Food Shack. It should be no farther than 100 yards away.

Cross City, Fl. (CTY)

Airport Restaurant 

They have a fine five burger restaurant serving breakfast, lunch, and dinner till 9:00 PM. The price of their meals represents the best value I have ever seen at any airport restaurant.

Crystal River (Crystal River Air Terminal - X31)

Dairy Queen 

Get a Blizzard and a burger at the Dairy Queen. It is within visual distance of the ramp. Later, hop a ride over to the Plantation for boat rentals and golfing. Don't rent a canoe if it's windy, get a flat bottom and go look for manatees on the river.

De Funiak Springs (De Funiak Springs Muni. - 54J)

McDonald's 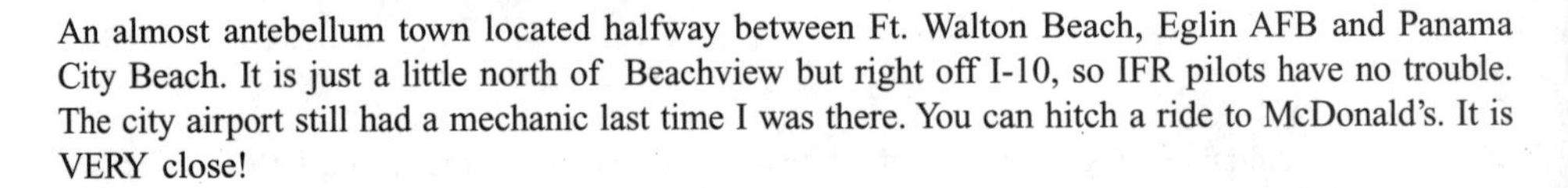

An almost antebellum town located halfway between Ft. Walton Beach, Eglin AFB and Panama City Beach. It is just a little north of Beachview but right off I-10, so IFR pilots have no trouble. The city airport still had a mechanic last time I was there. You can hitch a ride to McDonald's. It is VERY close!

Deland (Deland Muni. - DED)

SkyDive Deland 

The restaurant is on the field near SkyDive Deland. I don't remember its name. We were there about two weeks ago. I think the food was the best. The burgers cost about $6.00 but are worth it.

AMENDMENT

SkyDive Deland has a snack bar that serves delicious gourmet sandwiches and soups. The place is surrounded by friendly folk and colorful parachutes to boot!

Destin (Destin-Ft. Walton Beach - DTS)

Back Porch 

One of the best places in town for a fish sandwich is the Back Porch. It is located right on the Gulf of Mexico and about two miles from the airport. There is a courtesy car available or you may call a taxi. The cost is around $4.00 for a round trip. Truly a fine place to kick back and enjoy the sights as well as the food.

Everglade City (Everglades Airpark - X01)

Johnson Seafood 

The runway is 2,500 feet x 50 feet of asphalt. Taxi to the pay phone and call Johnson Seafood. Their number is permanently affixed to the booth. They'll pick ya' up in five minutes and take you to the docks, where you can buy fresh Florida stone crab claws, RIGHT OFF THE BOAT! They sell refreshments and dipping sauces too!

Bring a cooler and take some home with you. Just watch out for the gross weight! The airport is located on the edge of Everglades National Park. There are many attractions in Naples and Miami, not to mention the Keys, but that's a whole other burger.

Oyster House Restaurant 

Go through the bushes; there is a recently constructed gate marked to be open from 8:00 AM to

5:00 PM. It is on the approach end of runway 33 and across the street to the Oyster House Restaurant. You'll get the freshest seafood around, or a burger if you like. I recommend the grouper sandwich. It is delicious!

AMENDMENTS

I just returned from Everglades, FL. All the previous information on the restaurant is correct, but the food is way overpriced, tourist trap prices! I tried the grouper sandwich and had to call Jacques Cousteau just to locate the fish. The sandwich was $8.00!

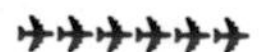

X01 was one of the most enjoyable Sunday afternoon flights I have ever taken. A friend and I decided to have lunch at the local seafood restaurant. I was served one of the biggest fish sandwiches I have ever seen. The food is not really overpriced at all. The service was great. The restaurant is about a 10-minute walk.

It's sad to say that all the great airports are small ones. I would mark this one down as a place to go for an afternoon or a full weekend. There's so much more to do than just eat. They have camping canoe rentals, and boat tours.

Ft. Lauderdale (Ft. Lauderdale Exec. - FXE)

Stephan's 

It's just a diner, but the folks are friendly, the girls are pretty and they have the best $102.69 burger available anywhere. The atmosphere there is the best. Once out the door and 10 steps to the left, you're on the flight line. Tell ground control that you want to go to Dolphin Flying Club right next to Banyan Air Service and you'll wind up right at their door.

Ft. Myers (Page Field - FMY)

Mike's Landing Bistro 

The restaurant is near the FBO, Ft. Myers Airways. Their prices are reasonable to slightly high. It is the right place for a business meeting, white linen on the tables.

In town there is the Edison Home. It is a must-see. You can go to the beaches of Ft. Myers and Sanibel Island. They're just a 30-45 minute drive. Ask for the courtesy car at the FBO.

Ft. Pierce (St. Lucie Cnty. - FPR)

Airport Cafe 

Ask ground to taxi you to the restaurant. Park right in front. Great burgers and service. Fine prices. Needs more window seats for us pilots.

Hollywood (N. Perry Airport - HWO)

Mayday's Restaurant 

Mayday's Restaurant has a wide selection for lunch and dinner. It is very, very popular with nonflying people. Often, there is a band on weekends. You'll find it on the south side of the field, just off RWY 9R.

Indiantown (Indiantown - X58)

Seminole Country Inn 

The Inn is a very good place to enjoy Sunday brunch. The staff is friendly and attentive, the food is great and the place itself is neat and clean. It is an actual hotel that has the Duke and Duchess of Edinburgh in its history.

A call to the FBO will get a van dispatched to pick you up in short order.

The hours are "normal business hours," but the brunch ends around 1500.

Key West (Key West Intl. - EYW)

Conch Flyer 

Yes, there is a restaurant at the Key West Airport, the Conch Flyer. Open 24 hours a day, like the Boca Chica Lounge on Stock Island.

AMENDMENT

After a trip to Key West from Fort Lauderdale, weather service advised returning asap due to incoming storms from the southwest. A quick meal from the airport restaurant, Conch Flyer, was a delicious and modestly priced surprise! Soup, fried conchs, atmosphere and great service made the first-time visit a pleasure.

Margaritaville 

Key West has it all, including polite Americans plus the best freak show every evening at Mallory Square. Park your plane, get a room for the night, go to Duval Street, stop by Fat Tuesday, get a cooler of pina colada, and head down to the old Navy Pier. At Mallory Square, watch the show and the sunset. If you have time, take a snorkeling trip or cruise. During the evening, have a beer at Sloppy Joe's and a Cheeseburger at Jimmy Buffet's Margaritaville. Don't buy too many t-shirts or you will not get off the ground.

La Belle (La Belle Municipal Airport - X14)

Flora and Ella's Restaurant 

Near the airport is Flora and Ella's Restaurant, known for the gigantic pies piled with meringue. The restaurant will pick you up or the FBO can arrange a ride. Located in a new Florida cracker-style building on State Road 80 in LaBelle, it is a landmark in the area for good eating. The local politicos congregate daily for breakfast and lunch. There is a nice gift shop in the lobby. Open daily except Sunday.

Lake City (Lake City - 31J)

Ken's Bar-B-Que

Lake City is definitely worth a stop to take in Ken's Bar-B-Que. The airport has a nonfederal control tower which operates part-time Mon.-Fri. Even though they have a tower, it is not Class D airspace. The runways are huge. For example, 10/28 is 8,002 feet x 150 feet. Like many Florida fields, it was once a WWII base. You may even see a ValueJet plane or two parked there, since Aero Corp. is one of the only remaining approved maintenance facilities.

Ken's Bar-B-Que is approximately one mile from the airport and just off the campus of Lake City Community College. This is the best barbeque I have had outside of Memphis. The prices are phenomenal. Wednesday is all-you-can-eat chicken day. The service is good and everyone is friendly. You will certainly get your fill. This definitely is worth a stop.

Lakeland (Lakeland Linder Field - LAL)

Tony's Airside

Try Tony's Airside at the base of Lakeland Linder Regional's tower. The food is good, service is fast and the prices are reasonable. You can taxi across to the Sun 'n Fun area and museum.

AMENDMENT

Easily the best fly-in restaurant in Florida: good ramp view, control tower right there, nice controllers. You may get a look at some show planes, WWII stuff on the ramp, who knows. Don't go there between 11:30 AM and 1:00 PM on Sundays. There will be a line of people trying to get in stretching out the door. The EAA Museum is across the field.

There be words said about pilots who overfly aviation museums.

Lake Wales (Chalet Suzanne Airport - X25)

Chalet Suzanne

Ask about the pilot's special. It generally includes dinner for two, a room for the night and breakfast.

This is an excellent restaurant, known for their Chalet Suzanne Soups, sold in many food-store gourmet sections. They also provide lodging in some unique rooms furnished with antiques. You can land at their private 2,450-foot turf runway which is adjacent to the restaurant or at Lake Wales four miles away. From Lake Wales call 813-676-6011 for the courtesy van.

AMENDMENT

Chalet Suzanne is very expensive, with prices running around $100.00 for dinner for two. The food is outstanding. They have a lobster Newburg-type dish that should not be missed.

Leesburg (Leesburg Muni. - LEE)

Capt. Bill's Seafood

Park at the far northeast end of field, not at the FBO. Walk toward the giant American flag, where you will find Capt. Bill's Seafood. You can't beat the prices or the quantity of food.

Marathon (Marathon - MTH)

Herbie's Bar and Grill

Marathon, Florida Keys, is halfway down the Florida Keys, 50 miles from Key West and 100 miles from Miami and 100 miles south of Naples, my home base. The FBO flight department is staffed by real friendly people and has a loaner car you can drive around town. I suggest Herbie's Bar and Grill. They have upgraded to concrete floors within the last 10 years but still have wooden picnic tables you share with other tourists. This is primarily a locals' hangout. Shrimp, conch and stone crab, when in season, are usually very fresh. All very reasonably priced at under $10.00. Don't be surprised if they are closed if the fishing offshore is good!

Miami (Kendall-Tamiami - TMB)

The Hangar Cafe

The restaurant at TMB is The Hangar Cafe, located in the Tac-Air building. The menu offers a good variety, including Key lime pie. Currently open Mon.-Sat.

Naples (Naples Muni. - APF)

Michelbob's

Michelbob's is about a 10-minute walk on the road running behind the airport. The FBO will gladly give you a ride there. Michelbob's claims to serve the best babyback ribs in Florida and they do so tongue-in-cheek. "Since baby pigs tend to rest on their right side, that side is tender before we cook it. We leave the left side to others. Ever heard the term 'leftovers'?" I think they are wrong with their claim. Their ribs are the best in the world!

Chrissy's

A great place for breakfast or lunch is Chrissy's, across from the Naples Airport. The FBO has a loaner car. Get the keys and ask for directions. It's in the building that says "Bingo" at the junction of Radio and Airport Roads. The service is fast, with a menu of homemade breakfast items, delicious banana nut bread, as well as soups and sandwiches and specialty dishes. Busy in season with mostly locals. No tourists!

New Smyrna Beach (New Smyrna Beach Muni. - 34J)

The Skyline Restaurant

It's not hamburgers, but a really nice fly-in restaurant located at New Smyrna Beach (34J). The Skyline is open for dinner only and is located on the airport just back from the ramp. The food and atmosphere are both great.

AMENDMENT

New Smyrna Beach Airport (34J) is south of Daytona Beach on US 1 near the inland waterway. It is just under the outer layer of DAB's Class C airspace. Ponce Inlet divides the southern shores of Daytona and the northern shores of New Smyrna and makes for some great sightseeing from the air. Keep an eye out for the lighthouse on the north side of the inlet.

The Skyline Restaurant is located between the visitor tie-downs and US 1 on the east side of the field. Beware, however, this is no burger joint. Although shorts are acceptable, it is not uncommon to see suits and ties. Weekend reservations are required! They are very busy.

Most items on the main menu are $20.00 and up; however, they do have an "early bird" insert with items priced between $10.00 and $20.00. They also had a lobster special when I was there, two for $19.99! All the food is excellent and you'll walk away happy. Service is friendly and talkative. Ask the maitre d', Doug Rudnick, for one of Jack's tables. He is a great waiter. If you really want, he'll probably serve you a hamburger.

Ocala (Ocala Muni. - OCF)

Pegasus Restaurant

A quiet airport with expensive food on the field.

AMENDMENTS

Ocala's airport restaurant is not all THAT expensive . It's an average airport coffee shop with pretty decent food. Prices are reasonable and about what you'd find anywhere else.

✈✈✈✈✈✈✈

Semiexpensive, good food, cute waitresses and nice people at the airport.

✈✈✈✈✈✈✈

I went to the Pegasus in Ocala for breakfast. The service was great. The waitress was gorgeous. The airport has a big long hunk of cement to land on. I will go again.

✈✈✈✈✈✈✈

I stopped in for the first time in late February. The food was very good, the service great and the prices in line. The waitress is a real knockout! I will make it a point to stop again.

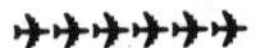

Not only is the food good, the service friendly and the prices reasonable, but the lineboy is a real stud muffin! For those of you not particularly interested in the server's knockout qualities, be sure to check out the lineboy!

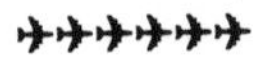

Pegasus is located on the field and within the Hawthorne FBO. It has nine tables with views overlooking the runway and additional seating at the counter. It is a simple restaurant and tavern. The tables are adorned with numerous old photographs of storm-damaged aircraft to remind you that Florida does have the occasional hurricane to disrupt the tranquil lives of the southern aviator.

Open every day from 8:00 AM - 11:00 AM for breakfast and 11:00 AM - 3:00 PM for lunch. It offers a respectable assortment of menu items. Breakfast selections range from omelets to pancakes. Lunch selections are a variety of hot and cold sandwiches, including burgers and soups. Expect to pay $4.00 - $5.00 for breakfast and $5.00 - $7.00 for lunch. They also offer beer and wine from $1.00 - $2.00 per glass.

RUNWAYS: 18-36: 6,906 feet x 150 feet of asphalt with pilot controlled lights. 8-26: 3,009 feet x 50 feet of asphalt.

Okeechobee (Okeechobee Cnty. - OBE)

Pizza Heaven 

Okeechobee will have a terminal building with restaurant within the next year. In the meantime if you're weathered in or just like checking out quiet airports for fun, there's Pizza Heaven 3 miles from the airport. They deliver to the airport, depending on the time of day, between 30-45 minutes. Use the payphone, 941-763-9929! Pizza, subs, pasta dinners, and salads are offered. There is a one-dollar delivery charge.

Orlando (Orlando Exec. - ORL)

Franchise Heaven 

Walk a half mile off the field to find a myriad (a lot, for you slow readers) of fast food and franchise food. Keep your head down and watch out for those jets going into Orlando International.

Pierson (Pierson Muni. - 2J8)

Carter's Country Kitchen 

Fly in to Pierson Muni 14 nautical miles northwest of Deland (DED). Carter's Country Kitchen is located approximately 150 yards, from the airport, with hamburgers, cheeseburgers, grilled cheese, patty melts, hot dogs, turkey, ham & cheese, roast beef, or club sandwiches. All priced from $1.95 to $5.35. Hours of operation, Monday through Saturday, 6:00 AM to 2:00 PM.

AMENDMENT

A new place just discovered is the grass airstrip at Pierson Airport (2J8). Carter's Country Kitchen is a block from the airfield. Open 6:00 AM - 2:00 PM and closed Sundays. Good home cooking, with the Wednesday special, fried chicken at $3.65, a favorite of locals and pilots who fly there especially for the chicken. Outstanding food and reasonable prices!

Punta Gorda (Charlotte City - PGD)

Airport Cafe 

Park your plane in front of the restaurant. It is open for breakfast and lunch. Owners change often but this is generally a clean spot with good prices and good food.

River Ranch (River Ranch Field - 2RR)

River Ranch Resort 

River Ranch recently changed ownership. Almost everything is back operational except for the marina and the marina restaurant. The main restaurant is open every day. A rodeo is held every Saturday night. A chuck wagon serves hamburgers, hot-dogs, etc., for the rodeo patrons. The saloon is open every Friday and Saturday night with a live band. Various style rooms, suites and cottages are available. They have three pools, numerous tennis courts, boat ramps, RV sites and much more.

AMENDMENT

The marina restaurant has reopened at River Ranch. It is now fully air conditioned and sports a new coat of paint.

St. Petersburg (Albert Whitted Field - SPG)

Coconuts 

Albert Whitted (SPG) in St. Petersburg, Florida, is a great place to fly in. The airport is directly on the water in downtown St. Pete. The FBO will provide free transportation, or it's a beautiful walk to the pier. There are several restaurants, ranging from very fancy (read expensive) to hamburger places. At the end of the pier there is a good-sized mall. It's a great place to take a few pictures, do a little shopping, and enjoy the beautiful scenery while you dine.

AMENDMENT

The restaurant on top of the pier is called Coconuts. There is a fantastic view of the airport itself. I love that place!

The Renaissance Vinoy Resort 

The Renaissance Vinoy Resort is a four-star old Florida hotel located very near Albert Whitted. The original hotel was built in 1925. It reopened in 1991 after a massive restoration. It has four restaurants that range from casual, Alfresco's for a sandwich or light meals, to elegant, the Terrace Room featuring a traditional American menu and a knock-your-socks-off Sunday brunch buffet, and Marchand's for superior Mediterranean cuisine. Be aware that it is more expensive than your average airport restaurant, $6.00 - $10.00 per person for lunch at Alfresco's or $35.00 for the Sunday brunch in the Terrace Room. If you call the hotel from SPG, they will send the hotel limousine to pick you up and will deliver you back to SPG after eating. If you want to rest overnight, they offer golf, tennis and spa. There are some neat museums nearby, like the Salvador Dali Museum and the Florida International Museum, which now has an Alexander the Great Exhibit and will have an exhibit on the H.M.S. Titanic later this year.

St. Petersburg (St. Petersburg Intl. - PIE)

94th Aero Squadron

The 94th Aero Squadron is located within walking distance from Jet Exec on the field at PIE. The food is good, the prices are moderate and the service is great. The Sunday brunch includes free champagne, although not recommended for the PIC. Great atmosphere with view of RWY 35L and 35R. Headsets are available at tableside for tower listening.

The Cuban Sandwich

If you are just passing through and want a good lunch, I recommend that you walk across the street from the airport. On the southwest side of PIE, near some T-hangars and where the UPS cargo planes are parked on the ramp, try the cuisine at The Cuban Sandwich shop. This is a little hole in the wall that's been there forever.

If you go there for lunch on the weekdays, you'll see three-piece suited lawyers right alongside truck drivers eating at the concrete picnic tables. The shop has heated and pressed Cuban sandwiches, Spanish bean soup, roast pork, deviled crab, stuffed potatoes, and other à la carte items. There are only a few places to sit outside, so you may want to order it to go. Back in the heyday of PIE, when People's Express was the hottest flight in, rent-a-car businesses popped up all over the place. Developers were flirting with The Cuban Sandwich shop owner to buy his property. He never sold. He knew the goldmine he had. People's Express folded and so did most of the land development in the area. The Cuban Sandwich shop still cranks them out at a couple of bucks per sandwich.

Sebring (Sebring Regional Airport - SEF)

Runway Cafe

The Runway Cafe, located in the terminal building of SEF, is right on the ramp. The owner, Mike Marlett, can be reached by phone at 941-655-5575 or Fax 941-655-5595. The hours of operation are Monday - Friday 10:00 AM - 6:00 PM, Saturday and Sunday 7:00 AM - 1:00 PM.

Smyrna Beach (New Smyrna Beach Muni. - 34J)

The Skyline Restaurant

It's not hamburgers, but a really nice fly-in restaurant is located at New Smyrna Beach (34J). The Skyline Restaurant is open for dinner only and is located on the airport just back from the ramp. The food and atmosphere are both great.

AMENDMENT

New Smyrna Beach Airport (34J) is south of Daytona Beach on US 1 near the inland waterway. It is just under the outer layer of DAB's Class C airspace. Ponce Inlet divides the southern shores of Daytona and the northern shores of New Smyrna and makes for some great sightseeing from the air. Keep an eye out for the lighthouse on the north side of the inlet.

The Skyline Restaurant is located between the visitor tie-downs and US 1 on the east side of the field. Beware, however, this is no burger joint. Although shorts are acceptable, it is not uncommon to see suits and ties. Weekend reservations are required! They are very busy.

Most items on the main menu are $20.00 and up; however, they do have an "early bird" insert with items priced between $10.00 and $20.00. They also had a lobster special when I was there: two for $19.99! All the food is excellent and you'll walk away happy.

Service is friendly and talkative. Ask the maitre d', Doug Rudnick, for Jack, as he is a great waiter. If you really want, he'll probably serve you a hamburger.

Tallahassee (Tallahassee Reg. - TLH)

Bagel, Bagel 

Bagel, Bagel has great sandwiches and is close to the airport. The FBO does have a crew car if you need it.

Titusville (Arthur Dunn Airpark - X21)

Dixie Crossroads 

The best fly-in eatery in the state of Florida? My vote is for Dixie Crossroads. They offer a full menu of terrific seafood, including all-you-can-eat rock shrimp for $16.95. The restaurant will even pay for your cab ride to and from the airpark! Go early, as the wait can often be more than two hours.

Titusville (Space Center Exec. - TIX)

Outer Marker Cafe 

The Outer Marker Cafe is at Space Center Executive Airport, formerly known as TICO Airport, in Titusville, Florida. You can park on the ramp right in front of the restaurant and eat inside in air conditioning or outside on a large screened porch. Overlook the runways and admire the often interesting assortment of aircraft parked in front. I once had to squeeze the Cessna 170 I flew in between a Stearman and a Luscombe. I was very careful! The prices are at the higher end of coffee shop, but the sandwiches are enormous and very good.

Venice (Venice Muni. - VNC)

Sharkie's Restaurant 

Park at FBO and walk a mile to the beach. Sharkie's Restaurant is a fun place to eat. Sunday brunch

is kind of expensive. They hand out prehistoric shark's teeth instead of breath mints at the end of dinner. Not great to use as tooth picks.
AMENDMENTS

Call Sharkie's, they will happily send the van to pick you up and later return you. Walk the beach first, and then try a drink on the rooftop benches for a view of the ocean and airport at the same time! They have live music sometimes on the back deck. Nonalcoholic beer is available at both bars, and virgin drinks at the inside bar, for the pilot in charge. A wonderful place to spend a few hours. The wait for a table can be long unless you go early on Friday or Saturday nights. The locals love it too!

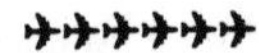

My wife, son and I recently made the flight from Tampa and had a super time. We could not have asked for the people at the restaurant to have been any nicer. They are used to fly-ins and really catered to us. We finished supper about 9:00 PM, and the shuttle was no longer running, so one of the waitresses piled us into her car for transport back to VNC. Really a great family place, you can walk to the beach and see a beautiful sunset.

Vero Beach (Vero Beach Muni. - VRB)

Cannon's

Park in front of the restaurant. Sit in a window seat and watch the flight school trainers hitting nose wheel first. Good food and reasonable prices.

AMENDMENT

Cannon's, located right on the field with a view of the planes is excellent and reasonable. They have a greenhouse-type area with plenty of window seats. Yes, you can park right in front of the restaurant and enjoy the view. Sunday brunch attracts the locals.

West Palm Beach (Lantana Muni. - LNA)

The Ark

Palm Beach County Park, commonly referred to as Lantana (LNA) on the airways, has a convenient restaurant called The Ark within walking distance from the FBO.

Huge, all-you-can-eat lunch buffet, fast service, clean and friendly. The FBO also has a small snack bar at this busy uncontrolled field.

Atlanta (Brown Fld. Fulton Cnty. - FTY)

Flight Deck Cafe

At Fulton County (FTY) in Atlanta there is the Flight Deck Cafe. They serve standard airport fare, breakfast through early dinner! I base there and I like it!

AMENDMENT

It is located right on the airport and the food is good. The only problem I have is the hours of operation: Monday - Friday 11:00 AM - 3:00 PM. I have been getting there before they open or after they close.

Atlanta (Peachtree-Dekalb - PDK)

57th Fighter Group

Peachtree-Dekalb (PDK) in Atlanta has the 57th Fighter Group. I've never eaten there, but I hear that it is very good.

AMENDMENTS

57th Fighter Group is a theme restaurant. It is just off the grounds of Peachtree-Dekalb Airport (PDK), an easy walk. Or park your plane near the end of runway 2 left and illegally jump the fence. But you didn't hear that from me.

The restaurant is decorated as a WWII officer's mess/French farmhouse, even though the 57th was based in North Africa. There are lots of photos and memorabilia on display. The atmosphere is truly fantastic and is best enjoyed on a warm spring evening when one can sit outside on the verandah and watch planes swoosh in on short final. The restaurant is just to the side of the end of runway 2 left. If you arrive earlier, you can arrange a ride in a restored PT-17 Boeing Stearman. I think they run about $80.00 and last more than 20 minutes.

It's good that the atmosphere is so unique, because the food is plebeian and a little expensive, $17.00+ entrees. The cuisine is uninspired country French. The best item on the menu by far is the seven-cheese beer soup. Guess how many varieties of scrap cheeses go into Velveeta — I kid you not. Its lineage notwithstanding, it's about the best cheese/beer soup I've ever had. The roast beef and chicken cordon bleu are adequate.

Part of the place is a bar/dance floor. It used to match the rest by playing WWII tunes, but it's pretty disco now. The crowd is from college on up to senior corporate pilots, actually, a pretty fun place, but out of sync with its surroundings.

Service is spotty; the waiters seem bored. The bar help is much better.

My wife and I, in our poor student days, frequented their excellent happy hours on Fridays. Their finger food included lots of breads, the cheese soup, and scrumptious Swedish meatballs — all for free! Alas, I do not think they continue with this great tradition, but the light fare still exists and is worth some money.

Based on other airport eateries, I give it 2.5 out of 5 burgers for dinner, 3.5 for brunch, Sundays, or lunch. The heavy food is adequate but not stellar, and overpriced. The atmosphere makes the place, and if you're eating light and out on the back porch, it really doesn't get much better!

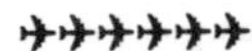

I agree with everything unkind that anyone can say about the 57th at PDK. The food is below average on their best days. The service is poor to nonexistent.

The Downwind

The Downwind is a good burger joint on the grounds of Peachtree-Dekalb Airport (PDK). This restaurant/bar is located in the old administration/tower building at midfield of the twin runways. Part of the place is indoors, but it has a large elevated deck that seats even more and has a great view of almost all airport operations. It can be reached from the street or the tie-down area, just in front of Epp's Air Service.

The building is a vintage WWII installation but has been refurbished in the last five years. The inside is now bright and cheery. The wall facing the runways is almost all glass. If the weather is poor, eating inside is not bad. If the sun's out, though, the deck is the place to be. Since it's elevated two stories, there are no bugs and usually a nice breeze.

The food is along the lines of a TGI Friday's or a Bennigan's. The staple is burger and fries, though the gyros and chicken sandwiches also are good. Upscale pub fare is probably an accurate description. Prices are reasonable, especially for an airport. You could probably come away full for lunch for about $6.00 – $8.00. Not the cheapest lunch in town, but hey, you're at an airport!

The service is fair to good. The waitresses seem a little overworked and not particularly into their jobs.

My friends and family often eat here as drive-up rather than fly-in. Given the amazing restaurants in Atlanta, this says a lot for the Downwind Cafe. It gets a 4 out of 5 burgers in my book. Could reach 4.5 with friendlier waiters/waitresses.

AMENDMENTS

The Downwind at the PDK airport is kind of a neat place to watch the planes come in. You can look over the runway from the deck of this vintage WWII building. The service is bad, and I have never had a good-tasting meal there. They have even messed up my salad, with brown lettuce. The burgers are fair to poor. The cost of food is nowhere near $6.00 a meal, unless you order just a burger and water. I work at the PDK Airport. When I want to have a drink and watch planes, I go to the Downwind. If I want something to eat, I go to any other place, including Micky D's, before the Downwind.

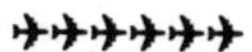

I think several others have missed some of the best food at the Downwind by treating it as just a hamburger joint. It is owned and run by a Greek family. It offers a real variety of wonderful Greek dishes, especially at night. Very good food at very reasonable prices. And I have always found the service just fine.

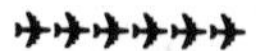

It's been a while since I frequented either of these restaurants, but I have to agree with most of the comments about them. My favorite of the two is the Downwind. Something that wasn't mentioned was how fresh the food is. I was told each hamburger is made when it is ordered and a turkey is cooked fresh daily. If this is true, it is pretty amazing. I've never had a bad meal or service. I'm sure if I went there frequently, I would. It happens at any restaurant.

Cornelia (Habersham County - AJR)

Brunch Bunch 

Fly to Habersham-Cornelia on Friday or Saturday evening or Sunday late morning. Park and walk on the road south, for a pleasant 5-10 minutes. Suddenly you and your flying pals will be at one of the all-around best buffet meals you will find anywhere. Nice crisp, clean salad bar, delicious vegetables, and first-class chicken and fish. I usually make a couple of passes here. There's more — include a fine dessert and you'll have enjoyed a first-class meal at a great place, and at an economical price.

AMENDMENT

There's a large variety on the Sunday brunch worthy of numerous trips. The hours are Friday and Saturday 5:00 PM - 9:30 PM, Sunday 11:30 AM - 2:30 PM. Phone: 706-776-1238

Dahlonega (Wimpys - 9A0)

Smith House 

The Smith House is known far and wide as a good family restaurant to visit when enjoying the Fall foliage in the north Georgia mountains. It is several miles from the airport, but they will send a van for you and are VERY used to doing so. Auto traffic in the mountains in the fall is awful, so lots of their customers actually do fly in. The ride in the van is great. Since you're not driving, you get to see the foliage.

The restaurant is one of those big family dining experiences. The food is so-so but reasonable in price, less than $8.00 for a BIG lunch. It's basic meat-and-potatoes. The three times I've been there, the service has been outstanding at this very friendly spot.

The town is worth visiting; it's all an easy walk. Georgia had a mini-gold rush in the late 1800s, and Dahlonega was its center. The gold dome on the courthouse, a great landmark for aerial navigation, is real gold lamé. You can pan for gold right on Main Street and do other goofy tourist stuff. Be sure to tell the restaurant van driver when you'll be back or ask if they are on a schedule. I've never had a problem coming or going when I liked.

Overall, I give the restaurant two burgers for the food, but I give the airport-town-restaurant experience a solid four out of five. Do it in the fall on a clear day.

Dawson (Dawson Muni. - 16J)

Varsity 

Down in south Georgia, on the north side of the Jacksonville sectional, is a town called Dawson (16J). There is a single runway and a real old-style FBO run by an older man and his wife. They appreciate cash for fuel, to keep prices low and because they don't take credit cards. The best part is they'll lend you their car so you can head up the road to the Varsity — no relation to the Athens and Atlanta Varsities — for a barbecue pork sandwich and Brunswick stew. OK, so it's not a hamburger, but what else is there, really, in south Georgia?

Irwinville (Crystal Lake Airpark - 3J8)

Splashtown 

A better place to fly in than Crystal Lake, GA, is hard to find. This place is a water theme park in mid-Georgia, just off I-75 east of Fitzgerald and north of Tifton. When you fly in, admission to the park is free. The food is typical amusement park snack bar variety but good if you're hungry. All the country cornfed Georgia women go there. Heck, it's the only beach for 250 miles. It appears on the JAX sectional.

Jefferson (Jackson Cnty. - 19A)

Porter House Restaurant 

This old converted school house is very near 19A. They have the BEST fried catfish and fried chicken you will ever eat. Both are served with all the fixin's. Call from the pay phone and they will pick you up and bring you back. The Porter House is open for Friday and Saturday dinner and Sunday lunch. If you tell them that Frank sent you, they will give you the best seat in the school house!

Jekyll Island (Jekyll Island - 09J)

The Old Hotel 

If you are interested in impressing someone with some great food, the Sunday morning brunch at

The Old Hotel, I think Radisson owns it now, is one of the classiest eating experiences you can hope to have. The hotel will send a van to the airport and return you when you're done. It was once a vacation retreat for the Rockefellers, Vanderbilts, etc. The food is spectacular. We recommend the sampler plate, but with the omelet station, salad buffet, and dessert bar, you can expect to leave the airport with a slightly different center of gravity. There is no FBO or fuel available at Jekyll, but Malcolm-McKinnon (SSI) offers a full-service FBO and is five minutes away.

Lawrenceville (Gwinnett Cnty. - LZU)

Flying Machines 

At LZU, Gwinnett, GA, on the Atlanta sectional, a fancy new place opened just in time for the Olympics: Flying Machines is on the south side of the airport.

AMENDMENTS

The Gwinnett County Aerodrome (LZU) now has an on-field restaurant called Flying Machines. It is owned by the 57th Fighter Group chain. They have great hamburgers and lots of other good stuff. They are open from 1100 - 2200 Monday through Saturday. Closed Sundays. The dinner menu looked excellent also. LZU is located in Lawrenceville, GA, close to Atlanta. Park your plane at the door and either eat inside or on the patio.

✈✈✈✈✈✈✈

The Flying Machines restaurant is closed. I talked to the old manager and he stated that the Olympic traffic did not pan out and they were overextended. No plans to reopen have been announced.

✈✈✈✈✈✈✈

I have been advised by the owner of the building that Flying Machines will be reopening 7/8/97. The menu will be somewhat more lunch-crowd friendly with better pricing.

✈✈✈✈✈✈✈

Great News! Flying Machines restaurant on the south side of the field has reopened. I had lunch there today and the food and service were excellent. They are open for lunch from 7:30 AM to 3:00 PM at present. After the establishment gets off the ground, so to speak, they will also include a dinner menu. Hours of operation will then be from 7:30 AM to 11:00 PM. Future plans call for tables outside on the deck, entertainment in the evenings, and they will offer biplane rides.

Macon (Smart Downtown - MAC)

Tipitina's Restaurant 

Tipitina's Restaurant at Herbert Smart Downtown (MAC) is located on the field; you can taxi right up to the front door. The food is strictly diner, and that's not bad. Everything from breakfast to dinner, just as if you were on the road. The burgers are good and lunch will set you back less than $6.00. Service the three times I've been there was good to very good, prompt and courteous.

The building overlooks the tie-down area and provides the best view of air operations on the field, short of the tower, but this is not very good. Most of the airport buildings are far removed from the runways.

Pine Mountain (Callaway Gardens - PIM)

Callaway Gardens 

This is a southern-type resort, with a beautiful garden. Pine Mountain (PIM) Airport is about a ten-minute drive.

If you call them on 122.8 about 10 minutes before you arrive, the resort will have a van ready and waiting. They will take you to the main hotel, where a great brunch is served.

It's a fun way to enjoy a morning or afternoon.

AMENDMENT

As of May 97 the PIM FBO is closed. Land anyway; there is a phone on a pole at the FBO which connects directly to Callaway Gardens Inn. They will send a van to pick you up for breakfast or brunch on Sundays at no charge.

Use caution on entering the traffic pattern. There are lots of planes on weekends.

Rome (Russell Fld. - RMG)

The Prop Stop 

The Prop Stop Deli at Russell Field (RMG) in Rome, Georgia, is very good. You taxi right up to the door and enjoy a good meal!

AMENDMENT

An excellent place to eat by any standard. This is so far above the usual airport fare that setting your course to the Prop Stop whenever you are near makes great sense. I have lunch there often and it is consistently wonderful!

Savannah (Savannah Intl. - SAV)

Hilliard's 

Fly into Savannah International, park at Signature (expensive) or Savannah Aviation (not quite as expensive). Either FBO will lend you a crew car. Ask for directions to Hilliard's. One is a mile or so away and another several miles away. Great seafood and amazingly inexpensive. I've not been disappointed .

Vidalia (VDI)

Vidalia Onion Factory 

Vidalia, GA (VDI), is in the heart of onion country. Ask Ken Nobles for the courtesy car and take the 1.5-mile or so trek to the Vidalia Onion Factory on US 280. The onion burger is tops. Nearby is Brewton-Parker College.

Benton Lee's Steakhouse 

If Ken Nobles will let you have the courtesy car long enough and draw you a map to Benton Lee's Steakhouse at Gray's Landing on the Altamaha River, you are in for a treat! Best be planning to spend the night in Vidalia. There are motels about a mile away, Shoney's and Holiday Inn Express are nearest. Benton Lee's does not advertise, but I did a piece on them for Fox Television and here's the scoop.

They do not serve on plates but TV trays. A small steak misses the sides of the TV tray by a couple of inches all around. The medium steak touches the sides of the tray, the large hangs OVER the sides of the tray. If you know of ANYBODY living around Vidalia, ask them to take you to Benton Lee's. As a result of the TV coverage, they get people from all over the world, but you won't see the place advertised ever. The steaks are well cut from selected Angus beef, and while you might have had better, more artistically cooked steaks, you will forget all about that!

Williamson (Peach State - 3GA7)

Barnstormer's Grill 

Nice front porch atmosphere and good chow.

AMENDMENT

This is a nice, family-run business, providing good food and good atmosphere! The screened porch has a fine gravel floor and is situated within 100 feet of the grass strip. Say hello to Shelby, the cute 4th grader-waitress.

Winder (Winder/Barrow - WDR)

Old Will's 

Summertime, Friday and Saturday evenings, STOP at Winder. Taxi to the east end of the field and park on the grass off the taxiway. Walk across the RR tracks and 200 yards ahead is excellent BBQ. The ribs and chopped are among the best commercial BBQ you'll find. Order the beans, not the Brunz stew, however. I'd come here even if it was not at the airfield!

AMENDMENT

Will's BBQ is now open for the summer and they are open for Saturday lunch too. Check for trains when crossing the track; it is quite active.

Peach Aviation 

Peach Aviation, formerly Northeast Georgia Regional Aviation, serves up hamburgers and hot dogs for $1 each on Saturdays and Sundays when they are pumping gas, 7:00 AM to 8:00 PM. They are the second building from the southwest end of runway 5.

Honolulu (Honolulu Intl. - HNL)

L&L BBQ

I know that this will make for a long flight for most of you, but if you are ever in the islands, then make sure you stop in at L&L BBQ. You can get the most ONO (delicious) burgers and food in the world. Order the teri chicken or hamburger steak. It comes with one scoop rice and one scoop Mac salad! Cheap and good; L&L is on Oahu, about five minutes out of Honolulu International. There are others all over the island as well as on other islands.

I flew out of HNL and Dillingham, which is a little sailplane and GA airport on the north shore. I grew up in Hawaii and am now attending the University of North Dakota majoring in commercial aviation and a minor in meteorology. It is much colder here. ALOHA!

Kaunakakai (Molokai - MKK)

Airport Café

I lived in Honolulu for a year and was able to rent planes from the Hickham-Wheeler Aero Club. I did make a few trips to Molokai and Lanai for lunch! First, I landed at Kalaupapa (LUP), the leopard colony, on Molokai and there wasn't anyone around so we took off and flew up to the main airport at Kaunakakai (MKK). We had a great hamburger there! And they sell the famous Molokai bread, which you can take with you!

Then, we flew to Lanai (LNY), where we caught a van that took us up to a neat little town in the pines at the 3,000-foot level. The air is clean and fresh, people are friendly, and there are a couple of small cafes where you can get lunch. It's difficult to find an FBO to rent planes, but on Lagoon Rd. near Honolulu International they have some planes for rent.

Athol (Silverwood - S62)

Lindy's

In northern Idaho there is a nice little airport called Silverwood. It is smack dab in the middle of a theme park. There is also an air show given in the early evening hours on a daily basis, as well as glider rides.

When you land, park on the grass at the northeast end of the runway. You'll have to pay $20.00 per adult; I can't remember the children's price. All rides and nearly all the entertainment is free. At first this seemed a bit steep, but we soon found out it's worth the money. To get more mileage out of your entrance fee and to avoid some of the crowd, get there early in the day. Aircraft parking wasn't a problem on the Saturday afternoon that we went. Auto parking was another story.

There is a well-maintained, full-featured campground across the street. The theme park is very clean, impressively built for what it is, and obviously aimed at good family entertainment. There is a restaurant called Lindy's on the grounds, but unfortunately you have enter the theme part to get there. So don't plan on eating unless you're planning to enjoy the theme park also. The food is very good, though.

We plan on going back next year. Silverwood is open from the beginning of May to the end of October. Plan on getting your avgas at nearby airports.

Big Creek (Big Creek - U60)

Big Creek Lodge

The Lodge offers two rustic buildings with food, trail rides, and lodging. If you like tall pine trees, spectacular mountain scenery, and a comfortable remote setting, this is it. I have only been to the Big Creek Lodge, but I have heard that the other one, Gillihan's Lodge, is just as nice. Breakfast at Big Creek is fantastic, and I have been told that lunch and dinner are just as good. Saturday mornings, you can usually find many backcountry flyers showing up for breakfast. It is recommended that you get reservations for meals and lodging, as both are small lodges (208-375-4921). This airstrip is in a mountainous canyon. Mountain flying experience is highly recommended. Be aware of high density altitude, turbulent air, and ever-changing weather conditions.

Burley (Burley Municipal - BYI)

The Cube

About 1/2 mile from BYI is The Cube, very reasonable prices and good food. Be sure to order a basket of their scones, delicious and unlike any I've tasted elsewhere.

Caldwell (Caldwell Intl. - EUL)

Cockpit Restaurant

The small on-field restaurant has hamburgers and a few sandwiches. Expect to do some hangar flying. The airstrip is adequate for any aircraft other than the "heavies."

Elk City (Elk City - S90)

Molly's

We spent a lot of this past summer "airplane camping" in the Idaho backcountry and enjoyed flying into Elk City for breakfast. There is a little restaurant in the main part of town that we always go to for a good meal. I think the name is Molly's, but I cannot be sure of that. If you go, it is the first building on the left side of the road as you walk up the hill into town. They serve a good breakfast and also have great pies. The General Store across the street is a good place to load up on supplies or snacks for the flight home.

Elk City Days is held the second weekend in August. It includes a parade, logging events, gymkana, cow pasture golf, dancing, and a big barbecue.

Elk River (Private)

Mother Lode Saloon & Steakhouse

Elk River is a private strip located at approximately N46-47.5 W116-10.5.

It is a north-south grass strip set in a nice valley. There are two restaurants within 200 yards of the strip and both have a country atmosphere. There are no services here but the town is very nice. There is also a lake within 1/4 mile that has swimming, fishing, etc. I have always had a good time when flying in there.

Emmett (Emmett - S78)

19th Hole

Emmett's FBO doesn't use golf carts to tow aircraft. The carts belong to the golf course whose office is located in what you would expect to be the terminal. This means that you can get breakfast or lunch at the same location during normal golfing hours. They have homemade hamburgers and sandwiches and a good selection of house-baked goodies, such as apple pie made with Emmett apples.

AMENDMENT

The food at the Emmett, ID (S78), restaurant is EXCELLENT. I suggest the "scramble" omelets. The burgers are the best I have tasted lately. Each one is individually prepared with anything you want on it. The lady who runs the place, Krista, and the others who work with her are great folks! Emmett is a regular stop on our Saturday morning "Dawn Patrol" trips.

Fairfield (Camas County - U86)

Country Inn

Fairfield, Idaho, is a small farming and ranching town whose airport has a hard gravel strip, suitable for any SEL or small MEL plane. Across the street is the Country Inn, which has the standard roadside restaurant fare, including some home-cooked entries and homemade soup. I find it to be a convenient stop on the way out of Sun Valley (SUN) or on the way home from a weekend.

Galena (Smiley Creek - U87)

Smiley Creek Lodge

This is a grass strip north of SUN. Across the road is a nice but plain restaurant. The surrounding country is some of the most astonishing in the USA. This is mostly a summer strip and is best suited for small SELs, but I have seen a Beech Starship put down there! As Idaho backcountry strips go, this is an easy one, with plenty of opportunity for go-arounds.

Hailey (Friedman Memorial - SUN)

Sun Valley Brewery

SUN is really the Friedman Memorial Airport in Hailey, where it is not unusual to find more jets on the transient ramp than recips. One time I counted 21 jets belonging to the Robins investment firm. The FBO's staff are very cordial no matter what you fly.

Tie down and walk a mile to Hailey. The choices include: (1) the Cafe at the Brewery (Sun Valley Brewery); (2) a pizza place; and (3) a small restaurant above a bookshop across the side street from the Cafe at the Brewery. All are on the east side of the road. All are worth the walk if you can't get a ride.

If you have time, drink a home brew at the Brewery and arrange to spend the night in Sun Valley, 11 miles north. The food there is world class.

The Western Cafe

If you can get to Ketchum from the Hailey airport, The Western Cafe is good for a solid American breakfast or lunch at a reasonable price. Their pies are also quite good. It's located on Main St.

To get from Hailey to Ketchum, I usually pack a pair of rollerblades and skate the 12 miles on the nearly perfect blacktop path which runs right past the Hailey airport. Most days you can also catch the Atkinson's shuttle from Hailey to Ketchum. Ask in the FBO. Last time I checked, the shuttle did not run on Sundays.

Idaho Falls (Fanning Field - IDA)

German

There is a German restaurant at the airline terminal. Pretty good, but I don't know how the staff tolerates the continuous Um Pa Pa music.

"Yes, but them Folkers was flying Messerschmitts."

Lewiston (Nez Perce Cnty. - LWS)

Airport Restaurant & Lounge 

A good ol' American restaurant serving steak and mashed potatoes and little else.

LWS is a favorite ramp check, so don't talk to strangers.

McCall (McCall - MYL)

Si Bueno 

If you want some Mexican, Si Bueno offers wonderful Mexican food. You cannot miss finding the building. It is the one with the tail of a Piper sticking out of the roof. Si Bueno is located across from the parking ramp. (208-634-2128)

The Pancake House 

McCall, Idaho offers a couple of great restaurants within a short walk of the airport. The Pancake House, 1/4 mile north of the airport, offers wonderful breakfasts that can satisfy even the hardiest of appetites. Try the "mountain man" breakfast which has a pork chop, eggs, hash browns, and pancakes. By the way, one pancake covers a plate. (208-634-5849)

If you need supplies, a supermarket is located across the street from the airport and offers some hunting, fishing, and camping gear.

Murphy (Murphy - 1U3)

Wagon Wheel Cafe 

Murphy, Idaho, has a population of about 10, a grocery store, 2 bars/restaurants, the obligatory house of ill-repute, and the county offices. They emphasize their desert atmosphere with a single parking meter on the gravel parking lot in front of the county building. Oh yes, they have three airstrips. Local pilots fly in for their "Murphyburgers." The original restaurant that had Murphyburgers changed hands, so you can take your choice of two establishments to get the burger.

Land from the west, wind permitting. The strip is slightly uphill. A wadi precedes the approach, so land with power and be prepared for downdrafts over the wadi. On takeoff, turn right to fly the wadi until you have altitude. The asphalt strip is used to train students in the early fundamentals of backcountry techniques.

Pocatello (Pocatello Muni. - PIH)

Blue Ribbon 

On the field at Pocatello, Idaho (PIH), the Blue Ribbon is open for breakfast, lunch, and dinner. Lunches run about $5.95 for anything from "gourmet burgers" to hot and cold salads. Dinners range from $9.95 to $15.95 with a diverse menu that will match any fine restaurant in the intermountain west. All items on the dinner menu have won first place, thus Blue Ribbon, at the Idaho State Fair competition. Potato dinner rolls you'd leave home for.

Sandpoint (SZT)

Pend Oreille Brewing Company 

In Sandpoint, Idaho (SZT), there is a new brewpub with the best burger ever! One-half-pound ground sirloin on a fresh roll with sautéed mushrooms and bleu cheese, and great Cajun fries!

The pub is downtown, two blocks from the lakeside public park. Free courtesy cars are available at the airport, so it's easy to get around. This is a lakeside tourist town with many good restaurants, shops, and community events occurring year round.

Don't forget the best $100 hamburger on the planet. It's at Pend Oreille Brewing Company!

AMENDMENT

The courtesy cars have minimal charges; I believe it's $5.00, plus 30 cents per mile. Still cheap, and four of us VariEze, LongEze drivers partook of the Pend Oreille Brewing Co. recently and wholeheartedly agree with the listing. The food and service were great. Alas, we could not bring ourselves to sample the Idaho Pale Ale and Hefeweitzen that are made on the premises.

Stanley (Stanley - 2U7)

Mountain Valley Lodge 

Another easy backcountry fly-in. In town there are at least two good restaurants offering hamburgers and other homemade goodies. This is almost always open summers and may also be open other times.

The airport slopes gently to the north. We are talking high-density altitude. Check with local pilots before takeoff if you're unfamiliar with mountain flying. If it is excessively hot, you need only wait until evening.

The Notty

The Notty is next to the Notty Ore House in Lower Stanley. You'll have to mooch a ride or borrow a car. It has really great garlic burgers for lunch and a limited but exquisitely done dinner menu. Chef Tom looks a bit like a hippy, not out of place for Stanley, and comes from a little further down river. Get Dia to play the piano for you at Stanley Air Taxi to stir up your appetite.

Warren (Warren Muni. - 3U1)

Winter Inn

Warren, Idaho, is a rustic, historical mining community that hasn't changed much in the last 100 years. The runway parallels main street and ends in "downtown." Winter Inn, about 100 yards away, is a hotel/restaurant/bar that offers breakfast, lunch, and dinner. I have only been there for lunch and had a real good hamburger and a BLT.

Hanging in front of the Inn is a plastic spotted owl. Warren prides itself on its annual "Spotted Owl Shoot" held over the Fourth of July weekend. This is in jest, of course, but remember, this is a mining and logging area.

Alton (St. Louis Regional - ALN)

Amelia's Restaurant 

They always have a great lunch special, and many times they have a buffet. The restaurant is on the airport, just north of the tower. There are outdoor picnic tables.

Bloomington (Bloomington Normal - BMI)

Arnie's

Good family dining as you sit 50 feet from where your plane is parked. This is a tower-controlled field with light traffic.

AMENDMENT

We tried to stop in and enjoy the fine dining at Arnie's. They would have nothing of it. It was 2:00 PM. They had a staff meeting at 2:30 and would not serve us. I was highly unhappy.

Cahokia (St. Louis Downtown Parks - CPS)

Airport Restaurant 

East St. Louis, IL, has a very nice restaurant located on the field. The food is good and the ramp parking next to the entrance is excellent.

Casey (Patchet Field - 1H8)

Richard's Farm 

The door is ALWAYS open at Casey Municipal, Kermit B. Patchett Field, with free access to a phone for local calls.

Richard's Farm is a wonderful diversion from the normal restaurant scene. Free transportation to this restored two story barn is provided by calling 217-932-5300.

Meals can be ordered from a menu, buffet style, or family style. Music is provided on the weekends. Expect a bit of a wait without a reservation and little or no waiting with a reservation.

A very pleasurable and unique dining experience. Real down-home folks! This is a don't-miss experience! A gift shop is on the premises selling local arts and crafts.

Champaign (Willard Airport - CMI)

Airport Restaurant 

The restaurant is in the main terminal building. General aviation parking is not allowed, but FlightStar, the FBO, is terrific and will give you a hop across the field. The food is good, the prices are very reasonable. They even have a little bar for cocktails for your passengers. All-in-all, a great place for lunch.

Aunt Sonya's 

Upon deplaneing at the FlightStar FBO, we were informed that the terminal where we had planned on having breakfast was closed due to the finding of a device that looked as if it could be a bomb. This was particularly unfortunate since WE WERE HERE FOR BREAKFAST AND BREAKFAST ONLY! As with all great FBOs, FlightStar volunteered a crew car without even being asked, suggested we drive to Aunt Sonya's, gave us a map and a pleasant "enjoy your breakfast."

Aunt Sonya's was a terrific suggestion! The line inside was long, but the locals suggested we wait since the line would move quickly, which it did. The service was very friendly, the food delicious, and the prices delightful. Two of us ate for less than $12.00, less tip. The others, well I didn't pay attention to their bill. Needless to say, we all had full, happy tummies!

I believe we all want to come back here some time and sample the food at the main Champaign terminal and on another mission to return to Aunt Sonya's.

Chicago (Meigs Field - CGX)

Gino's East Pizza 

Borrow a crew car and drive a few miles north up Michigan Avenue to Superior. Hang a right, travel one-half block to Gino's East Pizza, probably THE premier pizza joint in all of Chicago. Bring a pocket knife or some nail polish to leave your initials on the wall, as thousands before you already have.

Chicago (Midway - MDW)

White Castle 

This restaurant is a classic, known for its greasy onionburgers called Sliders. The Slider is so small that three of them are considered a snack. Here's the best part: this White Castle is right

on the southeast corner of Midway Airport, and it is so close to Midway Airport's runway 31C that aircraft are only 200 feet AGL as they pass overhead. Also, it is within walking distance of Monarch South, which is where transient GA airplanes may park.

Decatur (Decatur Muni. - DEC)

Main Hangar 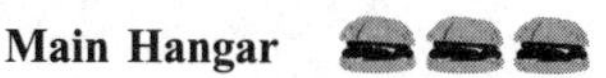

In Illinois try Decatur — right in the main terminal overlooking the controlled field. They offer a good breakfast, good lunch, good dinner and a good Sunday brunch. Park at the base of the tower. Many locals eat there too, so you know it's pretty good! Lots of model aircraft are displayed in the restaurant.

AMENDMENT

This morning my wife, a few friends, and I had breakfast at the Main Hangar in Decatur (DEC). It's about 115 nautical miles due west from Indianapolis, IN, just the right distance for a Mooney and Katana flying there in formation.

The restaurant is in the commercial terminal building, which we found to be locked from the ramp side. Once we found our way inside, we were treated to fast, friendly service, and very good food for an airport restaurant.

The breakfast buffet, served at 8:00 AM, cost $5.50 and had endless portions of biscuits and gravy, real hash browns, the best bacon and sausage anywhere, scrambled eggs, pancakes, and wonderful home-made waffles. You couldn't want anything else. My wife and I ate for $12.00, including tip. They even kept on bringing coffee and water after the bill was paid!

The airport is very nice, with three active runways, and is towered. The only drawback is a $3.00 landing fee for single-engine aircraft, more, I think, for twins.

I would encourage others to visit the Main Hangar for a weekend breakfast. They open at 6:00 AM Saturday and Sunday.

Hinckley (DeKalb County - 0C2)

Hot Rod Diner 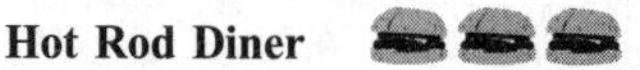

The Hot Rod Diner is right on the field. It is open summer weekends, serving hamburgers, hot dogs, deli sandwiches and salads. This is a fun Chicago-area grass strip with soaring and skydivers to watch.

Kankakee (Greater Kankakee - IKK)

Redwood Inn 

The Redwood Inn is two miles from Kankakee Airport (IKK). They have a good smorgasbord. Pickup service is available. Simply call IKK UNICOM.

Lacon (Marshall County - C75)

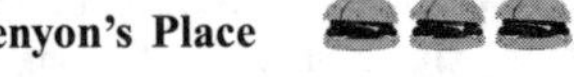

Kenyon's Place

In Lacon, about 1/2 mile from the airport, is a restaurant called Kenyon's Place. The food is excellent, and a free courtesy car is available at the airport for transportation. The runways have recently been resurfaced.

Lansing (Lansing Muni. - 3HA)

Little Diner

There is a small diner right on the airport at Lansing, Illinois, catering to the aviation community. I usually stop for a bite to eat and have a smoke when I am tired of doing touch-and-goes. The food is good, the service is extremely friendly, and they let you park your aircraft on a grass patch right in front of the diner.

Marion (Williamson Cnty. Regional - MWA)

Branson's

The restaurant is in the terminal building. The food is good and the prices are reasonable. They feature a smorgasbord at lunchtime.

Mattoon (Coles Cnty. Memorial - MTO)

Airport Steakhouse 

A decent airport restaurant I found is the Airport Steakhouse at MTO, Coles County Memorial Airport in Mattoon, Illinois.

AMENDMENTS

This restaurant is one of our EAA chapter's favorites. I defy anyone to find anywhere a sandwich as large as this restaurant's world famous "Elephant Ear." For about four bucks, this breaded and fried pork loin sandwich hangs over the edge of the nine-inch plate it is served on. They also have very nice dinner specials. I have even had Easter and Thanksgiving meals here, and they were very good.

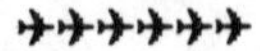

I have been to Matoon lots of times to eat and the previous report on the HUGE tenderloin sandwich is absolutely true. I've had barbecue, fish, and several other meals and have always been happy. The homemade pies are part of the attraction: peanut butter, chocolate, apple, cherry, really good with great crust.

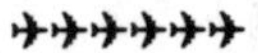

One of my favorite places to go is the restaurant at Coles County. Everything I have ever had there is good, and the prices are reasonable.

Morris (Morris Muni. - C09)

Runway 47 Restaurant 

Morris has one 4,000-foot runway. The restaurant has friendly people serving good food.

AMENDMENT

The name of the restaurant at the Morris, Illinois, airport is the Runway 47 Restaurant. They have a Saturday and Sunday morning breakfast buffet that is great. The homemade pastries are a necessity. The place is really busy, especially on Sunday, so a wait is sometimes inevitable. It is a fairly small spot in an old Quonset hut. The staff is very friendly!

Mount Vernon (Mount Vernon-Outland - MVN)

Midnight Landing Restaurant 

Mount Vernon has two paved runways. The restaurant is in the terminal building. I thought the food was good and reasonably priced.

AMENDMENT

The restaurant at the Mt. Vernon, IL, airport is not open for lunch on Saturdays and is closed on Sundays.

Plainfield (Clow Intl. - 1C5)

Airport Restaurant 

Good breakfasts on the field at Clow Airport (1C5).

AMENDMENT

The restaurant has been updated. The hamburgers are still the BEST you can find at any airport. This is a four burger spot in my book!

Quincy (Quincy Muni - UIN)

Airport Restaurant 

Good food and very reasonable prices. The control tower operates part-time.

Rantoul (National Aviation Center - 2I5)

The Fanmarker 

I recently flew into the airport at Rantoul, Illinois, and was delighted by what I found. Flightstar is the FBO. They have a great facility for pilots and multiple crew cars that they offer to all flyers free of charge. The airport is located on the now-closed Chanute Air Force Base. It has two excellent restaurants, The Caddyshack, located on the base golf course, and The Fanmarker, an aviation-themed restaurant housed in the old officers' club. After your meal, be sure to stop by the Air Force Museum adjacent to the Flightstar ramp. They have more than 20 interesting aircraft on display.

Seneca (Springbrook Marina - 1LL2)

Della Rose

This is a fun place to visit. First time in can be intimidating, but it really isn't as "character building" as it seems. If the approach calls for runway 36, keep your eyes open for rivertug boats. Some have tall structures, and the runway is only 20 yards or so from the river edge. A small walk will take you to the restaurant. We only had breakfast here and would rate it a full three. If you enjoy boats, you're in for an added treat.

Sparta (Sparta – SPT)

Rice and Fries

Sparta, Illinois, has one paved and one grass runway. Many restaurants are a short walk away. There is also a courtesy car available. I recommend Rice and Fries, a really good Chinese restaurant. McD's, Hardy's, and Pizza Hut are all visible from the pattern. All are a short walk.

Brandy's Bar & Grill

A nice spot just across the highway that runs by the airport in Sparta, Illinois, is Brandy's Bar & Grill. You can leave the courtesy car in the parking lot. You won't need it.

Wheeling (Palwaukee Airport – PWK)

94th Aero Squadron

Located on the field at Palwaukee airport (PWK), this is part of a national chain of aviation-themed restaurants. The food is always excellent and the service good. The dress ranges widely and casual is acceptable. Dinner price is around $15.00. The restaurant also features a Sunday morning breakfast buffet, with a twist: The earlier you arrive, the lower the cost! It is popular, even among nonaviators. If you go on Friday or Saturday night or Sunday morning, expect a wait.

The restaurant can be reached by van from either FBO, or on foot if you're willing to walk a bit. Unfortunately, there is no transient aircraft parking close.

Palwaukee airport is approximately eight miles north of O'Hare Airport. It is towered with two instrument approaches, ILS and VOR.

Bob Chin's

About 1/2 mile northwest on Milwaukee Road from Palwaukee Airport is a restaurant called Bob Chin's. It is an outstanding seafood place with an atmosphere that ranges from blue jeans to suits. The Mai Tai's are a must, although more than two and you will be flying without wings. Also try the crab legs.

Columbus (Columbus Muni. - BAK)

The Hangar V

The Hangar V restaurant serves breakfast and lunch. It is located in the terminal at Columbus Municipal (BAK). The lunch menu offers a wide range of items, including light meals and hamburgers. The food has always been good and the service very friendly. The large windows overlook the ramp, so you can keep and eye on your wings.

Though BAK is tower controlled, it is not very busy and has lots of tie-down space.

AMENDMENT

No longer open on Sundays, as expected. I ended up at George's, near HUF.

Elwood (Elwood - 3I1)

Roxie's Airport Inn

Roxie's Airport Inn offers simple home cooking. Great roast beef leads the menu! This airport was in its heyday back in the 50's and 60's. You can still see the signs proclaiming it as the world's first fly-in drive-in. I'm told that you could sit in your plane while being served by the car (plane)-hops! (317-552-6400)

There are two huge sod runways so wide that they look short, but they are both 2,300 feet by 300 feet.

French Lick Springs (French Lick - FRH)

French Lick Resort

For those flying around southern Indiana, a pleasant stop is French Lick Springs. At one time home to a casino, this resort is now run as a getaway location. French Lick Airport (FRH) is not exactly big or close to civilization. The gentlemen in the barn next to the strip will talk your arm off! The resort is always glad to send a van over for those visiting. This may tend to attract the middle-aged to older crowd. It is not exactly the greasy spoon associated with some airports.

Indianapolis (Eagle Creek Airpark - EYE)

Indy Rick's Boatyard Cafe 

On the west side of Eagle Creek (I14), situated on the reservoir, is Rick's Boatyard. Its big red roof is visible on downwind for RWY 3.

Opened in late summer of 1994, the menu runs from sandwiches to "California pizza". Pasta, seafood and steaks are thrown in. There is a complete beverage service. When the weather is good, the deck on the west side over the reservoir is great for outdoor dining.

AMENDMENTS

I just flew into Eagle Creek and ate at Rick's this past weekend and thought it was great. Terriffic food, excellent service, neat Sunday brunch menu, reasonable prices and wonderful views of the water, dam, and boat dock on the south end of the reservoir come together to make Rick's a winner! I had the French toast and it was really good. My friend ordered the ham and eggs. He got a big, thick piece of ham with Denver eggs and taters. We had to drain out some fuel to get him home. The locals say that in the summer, the wait can be long, but we got right in during the spring. Of course, there were no boathands in bikinis on the docks yet. I suspect (hope) that will be part of the summer allure.

I'll definitely go back, but I may have to leave my wife home and fly a low downwind to three over the lake.

Indy's Eagle Creek, "the Creek," as Indy approach sometimes calls it, has been changed by the Feds from I14 to EYE. Loran and GPS databases not updated since April 96 will need I14 while ATCC will need EYE to deliver you to a delicious meal and terrific setting.

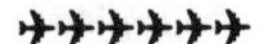

I went there with my wife, one of few places she will fly with me, and ate outside on the porch overlooking the reservoir with all the sailboats floating by. The catfish was wonderful. I even had to do a weight and balance calculation due to the large portions. Great place!

Lafayette (Purdue Univ. Airport - LAF)

Seattle Beanerie 

The Seattle Beanerie is about 3/4 mile north of the Purdue University Airport (LAF). It has specialty coffees and deli-type sandwiches. Call 317-746-3918.

Muncie (Delaware Cnty - MIE)

Vince's Restaurant 

PIREP

This is an excellent restaurant. Find it by the tie-down ramp in the operations building at the Delaware County Airport in Muncie, IN. It is a regionally known place to fly for a good meal. Reservations can be made by phoning 317-284-6364.

It is decorated with MAGNIFICENT model airplanes from the museum of the Academy of Model Aeronautics, now located here in Muncie. Many of these are antique models and many are of antique aircraft.

The restaurant has burger fare, but it also has an elegant menu and an excellent chef. It is one of two restaurants of choice for wining and dining prespective professional employees. Both restaurants are owned, in part, by Garfield, the cat, creator Jim Davis. Garfield memorabilia is available next door.

AMENDMENTS

The tower is once again in operation.

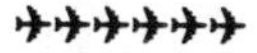

Early AM hours limited; always call before making plans!

Plymouth (Plymouth Muni. - C65)

Holiday Inn Restaurant 

The Plymouth Indiana Airport is located about 25 miles south of South Bend. The airport has a 3,500-foot east/west paved runway with plenty of parking at the west end. A paved diagonal walk-way connects the ramp to the nearby Holiday Inn. The hotel restaurant has a conventional dinner menu featuring steaks, prime rib and the usual. On Friday nights, the restaurant features a prime rib buffet, all-you-can-eat for $12.95. The prime rib is excellent, some of the best we've ever had. There is also a salad bar that includes shrimp and whitefish.

It is a two-minute walk back to your airplane. You can be in the air five minutes after you finish dinner. We found the service to be very friendly and highly recommend a visit to this midwestern restaurant.

AMENDMENT

We set flight for Plymouth, Indiana (C65), in search of breakfast. The restaurant is located very close to the ramp at the Holiday Inn. When you fly in, don't expect Unicom to tell you which runway is active.

Our first impression was not that good. We went on a Saturday morning at 8:00 AM local time. Service was VERY SLOW! We waited approximately 30 minutes from the time we ordered to the time breakfast was delivered. They appeared to have only two waiters working, or should that be overworking. There was a good selection on the menu, if you like eggs. The food itself was good, nothing to write home about, though! Big servings, however, except the daily special was a little thin. Two of us ate for $14.50, not including tip. Afterwards we sat out by the pool in the gazebo. Very nice! It was a great day! But then, isn't it so any day you can fly?

Schoop's

Located about a mile south of the airport is Schoop's. It is a 60's diner complete with all the trimmings — great hamburgers and a great environment. If you lived through the 50's and 60's you will love this place.

Portland (Portland Muni - PLD)

Richard's Restaurant

If it is dry, park at the east end of the taxiway and walk a couple hundred yards east, across the field to Richard's Restaurant, good homestyle food at very reasonable prices. If the field is wet or planted, plan on about a ¾ -mile walk.

Rochester (Fulton County - RCR)

Karen's

There is an excellent restaurant just across the road on the southeast side of the airport. Taxi over to the area from either end of the runway. Two restaurants there are very close to each other. The best one, Karen's, is on the southwest corner. Good food, generous portions, relaxed ambiance filled 2/3 with local folk and 1/3 by transient pilots. Be warned, Karen's is closed on Monday!

Gropp's

There are three restaurants almost literally on the field at Rochester, IN. Gropp's is on the south boundary. Gropp's opens at 11:00 AM, so no breakfast is served. During most of the year pilots can taxi to the restaurant.

Tony's

Tony's is one of three on-airport eateries at Rochester. Tony's is opened every day but Monday!

Terre Haute (Hulman Fld. - HUF)

George's

George's at the airport is located on Hulman Field (HUF), just east of Terre Haute, Indiana. Three runways are available and one is over 9,000 feet for you heavy drivers. George's is in the east end of the main terminal building and is a small, family-run place. George offers standard burger fare and great biscuits and gravy for Sunday brunch, but his real specialty is middle-eastern food. For a real taste treat, order the combination plate: humous, falafel, kibby, a small Greek salad, served with the freshest pita bread in town, for under $5. The best babahanouch west of Beirut. The hours are sometimes odd since George caters to the all-night freight service pilots, who keep late hours.

George's can be reached at 812-877-6777.

Be sure to look for the F-16s on the southeast end of the field. Watch out, they often do ops on the weekends and can be seen right out George's window doing landings and formation take-offs in groups of two and four.

Hulman approach, a full-time FAA tower with excellent, friendly controllers on 125.45, will keep you informed.

AMENDMENT

Good diner-style breakfast, made to order. Don't forget to ask about the specials. They are open seven days a week.

Valparaiso (Porter County - VPZ)

Strongbows Inn 

Strongbows is a short half-mile walk from Porter County Airport (VPZ) in Valparaiso. Great food! Pickup service available.

Amana (Amana - C11)

Ox Yoke

The finest fly-in eateries in Iowa are located in the Amana Colonies, south of Cedar Rapids in eastern Iowa. CID approach will even give you a vector.

Amana is a small grass strip, but plenty long for most singles. Many restaurants and shops are within easy walking distance of the field. The Amanas are an old German colony, and the restaurants reflect the homeland. The Ox Yoke is my favorite.

Anita (Anita Muni. - Y43)

The Redwood Steakhouse 

For The Redwood Steakhouse (712-762-4105) park on the north end. The walk involves a rustic trek up a slight embankment, across the railroad tracks, then a highway. Altogether, it is about one block.

Dubuque (Dubuque Regional - DBQ)

Cedars

We recently flew into Dubuque, Iowa. We intended to get a crew car and go into town to eat at Cedars. The crew cars were gone, but the FBO operators called Cedars. They were more than happy to pick us up at the airport and take us into town for lunch. They have an excellent menu and a very relaxed atmosphere for dining. The menu includes everything from steaks to sandwiches. Yes, they were more than willing to take us back to the airport. My wife and I usually take another couple with us to this particular restaurant. It was first suggested to us by the FBO at Dubuque Airport because the airport cafeteria was closed when we arrived. The food is terrific. I would rate it a three or four burgers. The phone number at the restaurant is 319-583-3211.

Keokuk (Keokuk Muni. - EOK)

Longenecker's Dinnerboat 

We recently flew to Keokuk on a full-moon-lighted night flight. The fine folks at Lindner Aviation let us use their courtesy Dodge minivan. They recommended the Hawkeye Restaurant, advertised as chef-owned, in the Pierce Hotel, which is listed on the National Historic Registry, and the Longenecker's Dinnerboat, adjacent to the Riverboat Museum. We drove six miles straight downtown to the river and Longenecker's. The bouillabaisse, salad bar, blackened/bronzed catfish, Creole shrimp, and service were all excellent.

Muscatine (Muscatine Muni. - MUT)

Muscatine Hotel 

Muscatine Muni. is my home airport. I fly all around the country, and Muscatine is not a gourmet haven, but if you're here for lunch or dinner, try the "old" Muscatine Hotel down on the river. They at least try. The FBO has a crew car and the hotel is about 15 minutes away.

Sully (Sully Muni. - 8C2)

Down Home Cafes 

Sully, Iowa, has a grass strip just two blocks south of the town square. There are a couple of "down home" cafes there which have decent food and a comfortable atmosphere. It is a nice change of pace to put the tires on grass.

Waterloo (Waterloo Muni. - ALO)

Airport Cafe 

Waterloo has a "canteen" type restaurant in a corner of the main terminal. You order at the pickup counter and eat with plastic forks on paper plates at square Formica tables. Choices are frozen pizza, tuna salad sandwiches, etc. Waterloo is Class C, with commuter service for several airlines. The food is mediocre to poor. If you are hungry, it works.

Beaumont (Beaumont - 7K9)

Summit House Country Inn & Restaurant 

This spot is nothing if not unique!

They have a fly-in grass strip, 2,400 feet. It is 44 miles from Wichita (ICT) on the 086 radial. The field elevation is 1,617 feet. Unicom is 122.9. After landing, taxi down a road, cross the main street and park by the ol' train water tower. It's pretty neat!

The Summit House Country Inn & Restaurant was established in 1879. If you like, phone ahead for more information. The Summit House number is 316-843-2422. I believe the hours of operation are Tuesday - Saturday 10:00 AM - 8:00 PM, Sunday 10:00 AM - 5:00 PM, closed Monday.

Enjoy! It's well worth the flight.

AMENDMENT

I had the "Summit House Special," a very good $100 burger! The Summit House is obviously the most successful establishment in Beaumont, a town of perhaps 400. Interesting aircraft parking spot just south of the Summit House, with a "skullpture" complete with a mangled propeller. Good fun!

Cottonwood Falls (Chase Cnty. - 9KO)

Grande Hotel

Land on this nice grass strip, then make the ¾-mile walk into town to the refurbed Grande Hotel. Nice atmosphere!

Council Grove (Council Grove Muni. - K63)

Hay's House 

Council Grove, Kansas, has a City Lake three miles from town. The airport is next to the lake. There are two short grass strips, maintained well. You will find a wind sock and painted barrels marking the strips.

The strip is on a well-traveled county road. Try the Hay's House in town for excellent meals — 112 W. Main (316-767-5911). They are closed Sunday night and Monday.

Garden City (Garden City Muni. - GCK)

The Flight Deck Restaurant 

The Flight Deck Restaurant is really nice. It brings a lot of town people out to eat. That says something, considering the airport is several miles out of town.

Goodland (Renner Field - GLD)

Butterfly Lake Cafe 

The Butterfly Cafe at Goodland, KS, is a nice stop. The people are very friendly and the food is good enough to attract the locals from Goodland for lunch. The FBO there even gives out goodie bags to student pilots when you top your tanks with a few gallons of gas on cross countries.

AMENDMENT

We agree that the Butterfly Lake Cafe is a good place to eat. We are from Goodland and eat there often.

Hutchinson (Hutchinson Muni. - HUT)

The Steakhouse 

The Steakhouse on Hutchinson Municipal Airport has EXCELLENT steaks and service. Buffet for lunch, too. It's not more than 50 steps from the GA parking ramp. You'll find them at N38-03.93 W097-51.64.

AMENDMENT

Hutchinson has a great restaurant right in the terminal building about 50 steps from the ramp area. It has just been beautifully remodeled and has a warm and friendly atmosphere. The decor has the kind of ambiance that pilots, as well as others, will enjoy. At the entrance above the doorway leading into the eating area is the tail section of a 1/2 scale biplane, positioned as though it has just flown through the wall.

You can eat smorgasbord-style and choose from several types of meats and vegetables, salads and wonderful desserts, or you can order from a complete menu. If you choose a steak from the menu, you can watch it being cooked on an indoor charcoal pit that is arranged to allow viewing from the dining area. The quality and taste of everything I have had to date have been superb and at reasonable prices.

The large glass windows allow a view of the three runways. Military trainers from Vance Air Force Base stop in frequently for food and fuel. The FBO has top-notch service, the kind where they actually smile when they tank you.

This restaurant stop has got to rate at least a five in anyone's book!

Marion (Marion Muni. - 43K)

Kingfisher's Inn

About a mile and a half from the 2,540 foot x 45 foot strip is a nice restaurant overlooking Marion County Lake. They will pick you up at the airport. The number is posted at the FBO. The place is called Kingfisher's Inn and is open Wednesday - Saturday 11:00 AM - 2:00 PM and 5:00 PM - 9:00 PM, Sundays 11:00 AM - 9:00 PM, closed Monday and Tuesday. We really enjoyed the atmosphere and food. The owners were very friendly. (316-382-3755)

AMENDMENTS

We have been to the Kingfisher's Inn several times and have always had a great time. They serve a variety of food, including homemade pizza and breads, seafood, Italian entrees, sandwiches, and charbroiled steaks. The telephone and restrooms are located in the FBO. (316-382-3755).

A 2,500-foot runway, paved and lighted with a beacon, greets you. It's just a short walk (1/2 mile) north to the Kingfisher's Inn on scenic Marion County Lake. My wife and I like to fly up for the evening buffet. Punch in N38-20.25 W96-59.50 on your GPS/Loran and enjoy a good meal with friendly country folk!

Paola (Miami County - K81)

Brownie's BBQ

Miami County Airport (K81), located in Paola, KS, is really wonderful. The restaurant is located right on the field. It's a great place for the $100.00 breakfast. After 10:00 AM, they serve wonderful barbecue. I'd recommend the barbecue beef sandwich with the baked beans.

AMENDMENT

K81 Miami County is known all around Kansas City for its food. Brownie's BBQ is probably the best in the state, right up there with Tony Roma's. We go there on weekends as much as possible. The service is great, the food is excellent, and the people are wonderful. Also, cheap fuel! We recommend it to anyone.

Sedan (61K)

Harmon House

I'm a big fan of Kansas' small towns. Sedan is a pretty one. The airport has two grass strips and is set north and above town. The downtown area seems almost unchanged from the '20s, with hardware stores, one "Ackerman's" in business over 110 years at its current location, and a spacious main street. Home of the Yellow Brick Road, an annual festival is held in May.

Harmon House is located downtown, about 1.7 miles from the airport. That is 1.7 easy DOWNHILL miles into town; a fairly "Himalayan" marathon to get back to 61K, however. I don't know if cabs are available.

The food was tasty, with the highlights being a prettily garnished old-fashioned pink lemonade and, naturally, dessert. Choose from cherry cobbler with lashings of cream, or à la mode and tapioca pudding. The menu was standard fare of steaks and sandwiches, with some down-home touches like meatloaf.

I give it three burgers, both for nice food and the unique Kansas atmosphere of this very pretty small town and the folks who live there.

Walking from the airport? Head to the highway through a heavily bullocked field, make a left, at the junction bear left, keeping the golf course on your left, then a right past the hospital. At the hospital, turn right to a sports field, then bushwhack one block farther west, then turn south to Main Street. The Harmon House restaurant is on the right.

Topeka (Billard Muni. - TOP)

Mentzer's Restaurant

I fly a Cardinal based at Billard and I have spent a small fortune at the restaurant. It's that good. Nothing fancy, just good homestyle cooking.

It is open 7:00 AM to 8:00 PM, except for Sunday and Monday, when it closes at 2:00 PM. They have a large menu and a popular buffet. You can taxi right to it, just east on the ramp from fuel service.

AMENDMENT

I flew there for the first time last week and was very pleasantly surprised. The restaurant abuts the ramp. We had a $4.99 all-you-can-eat dinner buffet that rates a 4 out of 5. I can't complain about that!

Wichita (Mid-Continent - ICT)

Ryan's Steak House

Just off the airport perimeter is a remarkable place called Ryan's Steak House. Any of the FBOs will drop you off there or lend you a crew car. It's only a three-minute trip. The restaurant is a cafeteria-style place, something like a Ponderosa steakhouse. Unlike a Ponderosa, the food is great!

You can feed an airplane load of people with sirloin steak, salad, and dessert for $6.00 each. Most of the people I take there say their steak is among the best, if not the best, they have ever had! The atmosphere and service are kind of Midwest rural, but the food is great. I often make ICT my fuel stop just to eat there.

Amarillo Grill 

Speaking as a Wichita resident, if you're going to take the trouble to go to Ryan's Steak House, I'd go to Amarillo Grill instead. Open for dinner, not lunch, it is about the same three minutes away and features mesquite grilled steaks that are excellent. Any FBO should be willing to take you there.

Bowling Green (Warren County - BWG)

Rafferty's

Bowling Green, Kentucky, is a great place to fly in and eat. Corporate Flight Management, the FBO, will take you anywhere in a courtesy car. Rafferty's restaurant is within walking distance. They offer a great selection of food, including steak, ribs, and chicken, all moderately priced. Expect a short wait for a table during dinner rush hour, especially on Friday and Saturday nights.

The Italian Oven

Also within walking distance is The Italian Oven, huge portions of Italian food. The wood-fired pizza is terrific.

Cadiz (Lake Barkley State Park - 1M9)

Lake Barkley State Park Lodge

This is another good Kentucky State Park. Call ahead on Unicom and the Lodge will send a van. A wide range of lodging and food services is offered at reasonable prices. Both breakfast and lunch are served buffet style. Located on LBL (Land Between the Lakes), the area offers great fishing, rental boats, etc. The van is usually prompt.

This is a great place to stop in for the day or weekend. This fine facility is part of the Kentucky park system. Van transportation is available to the golf course, skeet range, or marina. The Lodge offers many rooms with a lake view. Check out the tennis courts and weight room.

Not bad for a 4,800-foot paved site.

Falls of Rough (Rough River State Park - 213)

Rough River Lodge

Rough River State Park is a great place for breakfast, lunch, dinner, coffee or whatever. It is a wonderful state park facility, with the lodge and restaurant about a five-minute walk from the airport. It is frequented by aviators and aviation groups, including the annual Sport Aviation Weekend hosted by the EAA chapters of Kentucky. The strip is 3,200 feet (2-20) paved. Camping is available, and the facilities on the field include showers. Great place!

AMENDMENT

The restaurant at Falls of Rough State Park in KY is now open and has been for some time. It is indeed a wonderful place to eat. We do the 100-mile trip quite often for a tasty breakfast that is very inexpensive.

Gilbertsville (Kentucky Dam State Park - M34)

Patti's

My favorite place in Kentucky is at Grand Rivers. Patti's Restaurant is located only five miles from Kentucky Dam Airport (M34). This is an 1880 vintage restaurant. Their specialty is a two-inch-thick pork chop. I have eaten there many times and have never had a bad meal. They also feature homemade bread, served piping hot in a flower pot. Their pies are absolutely outstanding.

Reservations are needed for dinner, even during the week. They have a large local clientele, plus a large boater crowd from Lake Barkley. They have a zoo for the kids.

If you want to gain weight and have fun doing it, this is the place to go. They will come to the airport to pick you up, then deliver you back when you are finished. Great folks!

Louisville (Bowman Field - LOU)

Le Relais

For those of you who prefer slightly expensive, fancy French food, I highly recommend Le Relais at Bowman Field just southeast of downtown Louisville. They have a great wine selection.

Amendment

Le Relais for Bowman Field in Louisville, KY, is within 30 feet of my Champ.

Bearno's Little Sicily Pizzeria

One of Louisville's best restaurants is also at the airport. Bearno's Little Sicily Pizzeria is directly across the road, 100 feet from Le Relais. It has long been heralded as a favorite in Louisville.

Mazzoni's

Mazzoni's offers sandwiches and oysters. It is next door to Bearno's. I dare say that no other airport in the nation has as good a selection of quality restaurants.

Bowman Field Liquor Store 

Another great place for lunch only is the Bowman Field Liquor Store. It is directly across from the field. Jackie, the owner, has great plate lunches that truly fill you up.

Owensboro (Davis County - OWB)

Moonlight Bar-B-Que 

Fly into OWB and ask Unicom to call Moonlight BBQ. They will send over a van to pick you up. The food is great, and they offer a lunch buffet for around $6.00. Several kinds of barbecue, pork, beef and chicken are available, cooked to match any taste. The buffet includes a small dessert bar with ice cream and hot apple pie. Moonlight would qualify as a "hole in the wall," not much to look at but it has lots of atmosphere and good food. A down side for some might be the lack of a nonsmoking section. Well known throughout the region! When we opened our flight plan, FSS commented, "Enjoy the BBQ!" Even the ATC recognizes planes that frequent the Moonlight.

Franklinton (2R7)

Franklinton Golf Club 

Franklinton Golf Club is right across the runway from the terminal and within easy walking distance. The Sunday buffet is excellent, country cooking at its finest. You'll enjoy good views of the golf course and dock ponds while dining. The price is very reasonable.

Jennings (Jennings - 3R7)

Holiday Inn 

There is a Holiday Inn right on the field. You can park your plane within 100 yards of the hotel and restaurant. The food is reasonably priced. They have a decent selection.

AMENDMENT

I landed at Jennings on my way back from Tennessee. What a great place! There is a hard surface tie-down area adjacent to the hotel. A 200-foot walk brought me to the restaurant. They serve a really nice buffet for only $5.25. I was told that they have the buffet every day but Saturday and that it goes from 11:30 AM until 2:00 PM. Jennings is well worth the stop!

Lafayette (Lafayette – LFT)

Buns 

Lafayette is known for its fine restaurants and Cajun hospitality. However, the best hamburger place is just off the airport. It is too far to walk, but the FBOs will normally lend you a car or give you a ride. The restaurant is called Buns.

It is across the street from a car dealership. Buns has long been considered the best hamburger joint in Lafayette.

Lake Charles (Lake Charles Regional - LCH)

Player's Casino 

Call Player's Casino from the terminal for courtesy transportation to and from the casino. They have a very nice, reasonably priced buffet. If you are playing blackjack, mention to the pit boss that you're a pilot who flew in. He/she will almost always "comp" you for the buffet.

Marksville (Marksville Muni. - LA26)

Grand Casino Avoyelles 

The Grand Casino Avoyelles is located about 1/2 mile from the airport and has free shuttle service to the casino. They have a buffet for $9.95 that has more than 100 items on the menu and is one of the best I've ever had.

AMENDMENT

The buffet at Marksville LA26 is open 11:00 AM - 2:30 PM and 4:00 PM - 11:00 PM. They have a Cajun restaurant and a malt and burger grill in the casino. There is a new hotel with a pool attached to the casino. Bring a jacket; the building is kept very cold.

Monroe (MLU)

Airport Cafe 

Not much in the way of burgers but some of the best home-cooked plate lunches around. Believe it or not, it is the airport's cafe. If you're in town overnight, even the airport lounge gets going.

New Iberia (Acadiana Regional Airport - ARA)

FBO Buffet 

The FBO at New Iberia, Louisiana, Acadiana Regional Airport (ARA) has built a "hopping business" by selling all you can eat for $1.00 barbecue, crawfish and other Cajun specialties. On any given day, there can be dozens of military aircraft of all types stopping by for lunch.

New Orleans (Lakefront - NEW)

Flight Deck Restaurant 

The Flight Deck Restaurant is located by the old tower. Taxi right up to the door. The whole front is glass and overlooks the field. You can watch the aircraft take off and land while you dine. The food is excellent and the prices very reasonable.

Oakdale (Allen Parish - LA36)

Coushatta Casino 

Another one of my favorite destinations! A very nice pilot's lounge, featuring complimentary sodas and comfortable furniture. The FBO offers attended fuel from the trailer home next to the pad. Call the Grand Coushatta Casino for courtesy transportation to the casino. EXCELLENT buffet and fun.

Patterson (William's Memorial - PTN)

The Eagle's Nest 

I wanted to let you guys know about the burgers in Patterson, LA. They're at the Eagle's Nest on the field. There's even an aviation museum across the field. A great place to visit!

Shreveport (Shreveport Downtown Municipal Airport - DTN)

Peggy's Downtown Airport Cafe 

Peggy's Downtown Airport Cafe has been open for about a year in the main terminal building at the Shreveport Downtown Municipal Airport. Peggy's is open from 7:00 AM - 7:00 PM on all days. She serves a mean breakfast, lunch, dinner, and a great $100 hamburger. Give her a try! Her phone number is 318-221-2999.

Auburn (Aurburn/Lewiston Airport - LEW)

Pasquale's Restaurant 

Pasquale's Restaurant is a new restaurant at the Auburn/Lewiston Airport (LEW). The hours of operation are from 5:00 AM to 2:00 PM seven days a week. The restaurant is located in the airport terminal building and can be accessed from either the aircraft ramp or from the street for those who decide to drive. Pasquale Popola, an 8,000 hour ATP with several type ratings, welcomes all pilots to enjoy the very casual, friendly atmosphere he has created at Pasquale's.

AMENDMENT

Here's a great one for you! Auburn-Lewiston, Maine (LEW), finally has a restaurant on the field! The restaurant is run by Pasquale "Pat" Popola, a reformed commuter airline pilot. Pat still flies charter with Twin Cities Air Service, the FBO on the field, when time permits. The restaurant is clean, bright and cheery and the food is GREAT — all homemade & fresh. Pat specializes in an "all day breakfast" every Saturday. Check out the Italian hot dog - YUM!! The facility is small, but Pat will find a way to accommodate you.

Auburn/Lewiston is an uncontrolled airport with long, wide runways, precision and nonprecision approaches, no unusual obstructions or other worries, and a friendly, full-service FBO. This would be a great stop on an IFR cross country lesson — or for anyone looking for a nice airport with a great little restaurant on the field!

Augusta (Augusta - AUG)

Ardito's Restaurant 

Any stop in Augusta is worth a quick walk across the street to the main terminal building where Ardito's is located. Sit in a booth and look out on the main tarmac and watch the traffic come and go. Food's excellent, ample, and the prices are very reasonable. The owner and wait staff are really friendly. Closed Sundays and we can't figure out why, except the staff probably wants a day off, too.

Bar Harbor (Bar Harbor - BHB)

Lobster by the Pound

BHB lies on the mainland, just before the little causeway that takes motorists to Bar Harbor. The road outside the airport, a half-mile walk, is lined with lobster pounds, where steamed lobsters, fish and clam chowder, and seafood of all descriptions is available fresh. The restaurant right at the airport entrance is excellent; the others look good, too.

Belfast (Belfast Muni. - 98B)

Young's Lobster Pound

Belfast is a quaint harbor town that seems to be overlooked by most tourists. It doesn't get the attention of Camden or Bar Harbor and it doesn't get the crowds either. Belfast has a fine airport that is about a one-mile walk from downtown. Walk out the airport road, turn left at the T, cross Route 1, a main road, and continue into town. It will be obvious where the center of Belfast is. About one mile down you have to take a right and go downhill a few blocks. There are a number of restaurants that have fine food. Our favorite is Young's Lobster Pound. It is a long walk from the airport. There are taxis in Belfast, and Young's is worth the trip. At Young's, you can order lobster, steamers, and corn and have it cooked in salt water. You can then eat it inside or outside on tables on the banks of the north shore of Belfast harbor. The prices are right, the food fresh, and the view serene. Young's has a year-round seafood market on the airport road that will pack lobster and other seafood for your trip out. The market is just a couple hundred yards from the airport.

90 Main

The address is the same as the name, 90 Main. They serve downeast regional foods prepared in an eclectic, gourmet style. They have their own bakery and serve unique and downright decadent sweets for breakfast or after meals.

Darby's

Darby's on High Street next to the movie theater is within view of 90 Main. It is more of a pub atmosphere, with a tin ceiling and an old bar. The good food is more bistro type. The mood is friendly and comfortable. A favorite of locals!

Weathervane

Weathervane on lower Main Street at the public landing has the best views and is very popular. Even though this is a seafood restaurant chain, it serves a wide variety of excellent food at reasonable prices. They have outside seating and large windows that allow you to feel that you are eating at the banks of the harbor, and you are.

Bethel (Col. Dyke Field - 0B1)

The Sunday River Brew Pub 

The Sunday River Brew Pub is located approximately one mile NE of the Bethel, Maine, airport. The restaurant serves food typical of generic brew pubs, i.e., good burgers, ribs, salads, etc. Of course, they also have the required excellent selection of various beers brewed on site. Remember, not while flying. It can be walked, but I would recommend calling Unicom 5-10 miles out and asking them to call a cab. It will be waiting for you when you land. The fare is only $2.00 - $3.00 per person. If you want to sample the beer, stay at any of the local B&Bs or hotels. For the downhill adrenaline junkies, Sunday River ski area is a $9.00 cab ride away.

Biddeford, Maine (Biddeford Muni. - B19)

The Dry Dock 

Biddeford, Maine, is less than 10 miles south of Portland. The Dry Dock is a five-minute ride in a Lincoln Continental, provided as the free courtesy car by Biddeford Muni. The specialty is fresh seafood. Try a cup of chowder, $2.00. The seafood basket includes a generous helping of several types of seafood for $6.95. The largest strawberry shortcake I've ever had goes for under $3.00. I suggest you divide this with a partner as the portions are too large for an individual.

Millinocket (Millinocket - MLT)

The Hotel Terrace 

Head north towards Mt. Katahdin and land at Millinocket (MLT). Although the airport has a grill that occasionally functions, hoof your way a short distance to The Hotel Terrace for lunch. Be sure to say hi to the proprietors, the Bisques. The food is cheap, plentiful, and if you are weathered in, you are at the hotel.

AMENDMENT

Just a note to say how good The Hotel Terrace is at Millinocket, Maine. Last Friday their menu featured a buffet including a great fish chowder, green salad, mashed potatoes, fried fish, meatloaf, veggies and dessert for $3.95. The atmosphere is great too, a big stone fireplace and clean spacious surroundings.

Portland (Portland Intl. – PWM)

Ricetta's Brick Oven Pizzeria 

This is the best gourmet pizza we have ever eaten. It's better than the $100 hamburger. Portland has an FBO, Northeast Air at 207-744-6318, they have a courtesy van available and a free map of the area. Ricetta's is a 5-10 minute ride from the FBO — by the Maine Mall — at the junction of Gorham Road and Philbrook. Lunchtime is an all-you-can-eat buffet of soup, salad and gourmet pizza, including a dessert pizza. I promise you'll love it! The restaurant's phone is 207-775-7400. I suggest you arrive at Ricetta's before noon to beat the local rush.
Mangiamo!

Rockland (Knox Country Regional - RKD)

Owl's Head Transportation Museum 

The Owl's Head Transportation Museum is on the field. It has a concession stand which serves burgers! There are many airshows here throughout the summer.

Sanford (Sanford Municipal - SFM)

Rudy's Cockpit Cafe 

Rudy's Cockpit Cafe at the Sanford Municipal Airport has got to be it. Rudy opened it after leaving the Jordan House B&B where he was the chef for many years. The cafe is located in the main terminal, and is handicap accessible. There's plenty of parking on the ramp, but it can get crowded on a weekend. Two FBOs provide the fuel, and the prices are some of the cheapest around.

The fare at Rudy's is simple and ample. The motif is WWII with plenty of interesting pictures and memorabilia. They make the usually short wait seem almost instant. The cafe is small, about 20 tables, so there may be a wait on nice summer weekends, but that's a small price to pay for good food. The menu consists of the usual cup of Joe, eggs, toast and pancakes. Be sure to check out the specials, and do check out the fresh-baked rolls and muffins. These change daily and tend to sell out quickly.

AMENDMENTS

I'll second the four hamburger rating. Sanford is what I compare all the other New England fly-in restaurants to, and so far they all come up short. Good food, authentic, not chintzy, decor, good service, good runways, and low prices on food and avgas. For me this is the real Maine.

One little detail, the Cockpit Cafe closes at 2:00 PM at least during the week.

Turner (Twitchels - 3B5)

Lobster at the Mini Mart 

If you're going up through central Maine and need a fuel stop, you can land just north of Lewiston at Twitchels. This is also a seaplane base. A two-minute walk across the road takes you to a Mini-Mart that has good coffee and lobster sandwiches. Not much in atmosphere, but OK food.

The 2,300-foot strip has trees at both ends so drag it in a little. This is kind of an old-school family airport.

Chickadee Restaurant 

Located just across the highway from Twitchels is the Chickadee Restaurant. Excellent local dining with many seafood specials. Watch your gross weight on takeoff after a meal here!

Annapolis (Lee Airport - ANP)

Hayman's Restaurant 

Try Hayman's Restaurant near Lee Airport (ANP). Lee is located just south of Annapolis, MD. It has a 2,000-foot runway and plenty of transient parking at the SE end of the field.

Hayman's is a short walk, about two or three blocks south along Route 2. Their claim to fame is delicious crab dishes. Try a softcrab sandwich or eat 'em by the dozen. If you want to sample the finest that the Chesapeake Bay has to offer, give Hayman's a try. (410-798-9877)

When you land, walk over to Chesapeake Air, the repair facility. Ask for directions.

Adam's Rib House 

Adam's Rib House is a little more than 1/2 mile down Mayo Rd. from the airport. (410-267-0064)

Baltimore (Martin State - MTN)

Decoy's Restaurant 

If you fly into Martin State Airport, get a ride in the airport courtesy van to Decoy's Restaurant. It's about a five-minute ride from the general aviation parking area. They serve burgers and Maryland-style seafood. Good lunch or dinner stop!

AMENDMENT

The Decoy is at the waterfront and you can eat on a deck over the water. Boats tie up at their pier! Very good atmosphere!

Cambridge (Cambridge - CGE)

The Runway Restaurant 

Next time you're passing near Cambridge, MD, give The Runway Restaurant a try. It's located at the NW end of the Cambridge/Dorchester (CGE) Airport in part of a hangar. They open and close early, so plan on breakfast or an early lunch. The food is abundant and inexpensive. A Sunday morning breakfast is my favorite.

AMENDMENT

The Runway Restaurant is a small cafe located on the airport, open from 6:00 AM to 2:00 PM daily except Sunday, which is 7:00 AM to 2:00 PM. Typical diner fare, including sandwiches, for a reasonable price. (410-221-0883)

College Park (College Park - CGS)

The 94th Aero Squadron 

Consider the 94th Aero Squadron at College Park (CGS). WWI decor, slightly upscale, this place is a short, slightly awkward walk from the terminal building. Good food served with a great view of the runway. (301-699-9400)

AMENDMENTS

The 94th Aero Squadron Restaurant is a solid on-airport restaurant. One of a large chain. Always good!

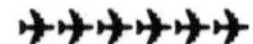

94th Aero Squadron is worth the trip. A bit pricey, but the food and service are good, and the WWI aviation decor is interesting. It overlooks CGS runway and can be crowded at lunch, especially during Sunday brunch, but it's worth the wait. Call ahead for reservations (301-699-9400). Happy hour brings out large crowds, but dinner is good and there is usually no wait for a table in the evening, since their main business comes from the nearby industrial park during the day. You can walk there from the transient parking ramp or to the nearby (two-block walk) METRO subway station, with frequent trains to metro Washington, D.C.

Cumberland (Greater Cumberland Regional - CBE)

Hagenbuch's Restaurant 

The FBO can sometimes offer a courtesy ride. (304-738-9907)

The terminal for the Western Maryland Scenic Railroad is three miles from the airport. The 32-mile, 3-hour round trip from Cumberland to Frostburg passes through some of the most scenic countryside in the region, including Cumberland Narrows, Helmstetter's Horseshoe Curve, a tunnel, and mountain scenery, all along a 1,300-foot ascent. It operates all months except January, February, and March. Call 1-800-TRAIN-50 or 301-759-4400 for details.

Easton (Easton - ESN)

Touchdown Luncheonette 

Great restaurant! Have a super breakfast for about $2.85 or lunch for approximately $4.95. It is super busy on Saturdays and Sundays.

This restaurant frequently fills the entire ramp! See you there! (410-822-8560)

AMENDMENTS

The Easton Airport Cafe in Easton offers good food, good prices, and a decent view of the runways. My personal favorite was the hot cider they serve, a great winter drink. Sandwiches were good, and the staff welcomes folks who just want to sit, drink some coffee, and watch the planes go by. Lots of families came and went while we waited.

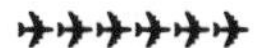

The Easton airport cafe is all that the other folks have said it is. However, you need to experience their seafood gumbo, a real taste treat. Good food, good service and a great way to do some quality hangar/cafe flying.

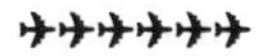

The cafe at the airport has very good Maryland crab soup. The burgers and fries are both homemade. The staff is friendly, and there is generally some very good hangar flying. Check it out, great view of the main runway!

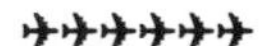

They are now under new management, and as far as I am concerned they serve the best hamburgers I have ever had. It has to be prime chuck they are using.

Gaithersburg (Gaithersburg - GAI)

Montgomery County Airport Cafe 

Try Montgomery County Airport (GAI) for a good meal. GAI is located near Gaithersburg, MD. No tower, but lots of traffic! Keep your eyes open and your radio turned up.

The restaurant is on the field and promises good food for most palates. They make a good hamburger for those who insist on tradition.

Give 'em a try, but not on a beautiful Sunday afternoon. It'll be difficult to park your plane and even harder to park your tail in a seat at the restaurant.

AMENDMENT

The Airport Cafe is located in the terminal building, open daily from 6:30 AM to 8:00 PM, for breakfast, lunch and light dinners. The food is good and prices reasonable. It is a popular place. The owner plans to add a deck on the flightline side in the future, for additional summer month seating.

Hagerstown (Washington County Reg. - HGR)

Nick's Airport Inn

Located across the parking lot and within about 100 steps of the local FBO, Aerosmith, on the GA side of Washington County Regional Airport (HGR) is Nick's Airport Inn. This has been a Maryland landmark for many years. Although a place of fine dining, Nick's has a special affection (or is that "affectation") for pilots and their passengers, routinely placing them boldly within sight of the normal gentry who frequent the establishment. They do make a great hamburger. Their best sandwich is the crab cake Maryland-style. It is made with lots of rich, sweet , lump crabmeat, just enough filler to keep it on a bun and then golden broiled to perfection. This is all served with linen, silver, and big smiles. If your weight and balance can stand it, they also serve a first-rate prime rib from a full menu with many other good seafood selections. Prices are not Denny's specials but are appropriate for the food and service provided. (301-733-8560)

By the way, when your park at the FBO, stop in and say hi to a great guy, Smitty, of Aerosmith, and kid him about what all those crab cakes have done to his waistline.

AMENDMENTS

Nick's Airport Inn is on the airport. It is recommended by many pilots in the area. The restaurant is located just off the northeast corner of the airport, behind Aerosmith's large blue hangar under the Phillips 66 sign. Access to the restaurant is through the FBO. Closed Sunday.

✈✈✈✈✈✈✈

One of the best airport-accessible restaurants I have found in 23 years of flying. Just a few yards walk from the FBO and ramp. It is excellent in all respects, yet reasonably priced. Great fried oysters!

✈✈✈✈✈✈✈

I flew to HGR three weeks ago to take advantage of the great food at the Amish grocery store. We got there just in time, because they were having a "going out of business" sale. Don't know when it will reopen or exactly what will be sold there when it does. When we were there, everything in the store was for sale, including butcher knives, saws, etc., so it's not just a matter of new management. If a new store does open, it will be a completely different enterprise. Better plan on eating at Nicks...then take a walk up the hill to check out the status of the Amish grocery store!

Tony's Italian Restaurant 

Tony's Italian Restaurant is also a great place to eat at Hagerstown, MD, Washington Co. Regional Airport (HGR). It is about a 300-yard walk from Alphin's FBO, or you can get a ride over from the FBO at Aerosmith. The food is great and prices are extremely reasonable. By the way, if you do what we did the last time over there and close the place up after FBO hours, the management of Tony's will give you a lift back to the airport tie-down area. This place represents a good five meatball subs rating.

Ocean City (Ocean City - N80)

Captain's Galley 

Ocean City, MD (N80), is, as we all know, a great place for all types of food. One of my favorites is Captain's Galley located at the harbor. It features one of the best crab cakes around.

Ocean Club 

You could go into town to the Ocean Club on 49th Street for a great prime rib — huge hunk of meat. Even the ladies portion should be plenty for anyone. Take the shuttle from the airport into town for a couple of bucks or share a cab with others. Once in town the Ocean City bus is only 75 cents to anywhere.

AMENDMENT

I flew into Ocean City, only to discover that the shuttle between the airport and the boardwalk is now closed. The receptionist in the FBO told me that this was permanent and not just a "closed for the season" situation.

She did suggest that we could get to the boardwalk if we wanted to rent a car.

Waterman's 

I have flown into Ocean City on several occasions and have always taken a cab wherever I wanted to go. To the beach only cost a few dollars and he came back to pick us up at exactly the agreed-upon time. Other times I have taken a cab to Waterman's restaurant for steamed crabs, all you can eat, and corn on the cob. Maryland is justly famous for its steamed crabs, and Waterman's will give you stories to tell your grandkids.

Salisbury (Wicomico Co. Regional - SBY)

Mike & Dave's Grill 

A new operator took over the SBY terminal restaurant in the summer of '97. Called Mike & Dave's Grill, they are open for breakfast and lunch and offer a breakfast and sandwich menu. They have very early hours for breakfast to catch the Piedmont Airlines morning shift change. They also advertise a Happy Hour from 4:00 PM - 6:00 PM, with draft and bottled beer available.

Stevensville (Kentmorr - 3W3)

Kentmorr Marina Restaurant 

Kentmorr Marina Restaurant at Kentmorr (3W3) is near Stevensville, MD. It's out on Kent Island in the Chesapeake Bay. The strip is 2,000 feet of turf. Walk about an eighth of a mile south of the field along the shore to the marina. You'll easily spot the restaurant. They serve a pretty good crab cake sandwich. (410-643-2263)

AMENDMENTS

The black angus beef sandwiches could make for a high-density takeoff on the best of days. You'll have a hard time finding a heavier beef sandwich!

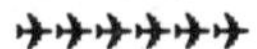

Kentmorr Airpark is a grass strip, public-use residential airpark located on Kent Island. It is about five miles south of Bay Bridge Airport. Great crab cakes and black angus beef sandwiches at nearby Kentmorr Marina Restaurant (410-643-2263). Just park on the north side of the runway, at the Chesapeake Bay end, opposite the homes and hangars, and walk a block to the marina. The food is excellent. This makes a good fly-in lunch spot, with a great view of the water and all the marine traffic in the bay.

Stevensville (Bay Bridge - W29)

Hemingway's

For a great view of boats, bridges and airplanes, try Hemingway's for lunch, dinner or Sunday brunch. Its located about half a mile northwest of the field. You can walk across the runway or give them a call from the FBO and they will come and get you. Mostly seafood with some landfood, busy on weekends in the summer, lots of boat people.

AMENDMENTS

When going to Hemingway's, you might want to call ahead to make sure there is not a boating event. I was going there one Sunday morning in the spring and found there was a National Boat Racing event that was going to attract 30,000 people to the area. Also, the Annapolis Boat Show is in October, and I am sure it attracts a lot of folks so I would call ahead if you expect a quiet day.

W29 is at the Bay Bridge in Maryland near Annapolis. It's a short walk to Hemingway's from the airport. It's on the Potomac waterfront, and the view is nice. The food was very good and the pricing fair. Seafood is the big deal here.

Hemingway's, 410-643-CRAB, is in easy walking distance. Follow the gravel road to the water and then the walkway along the water to the last building on the path. A casual dining spot on the banks of the Chesapeake Bay with an unobstructed view of the sunset near the base of the Chesapeake Bay Bridge on Kent Island. You may sit inside or on the deck. Prices are moderate $25.00 - $35.00 per person for dinner, and the food is excellent. Their specialty is crab. Reservations are suggested.

Kent Manor Inn

Want to fly somewhere for a meal that's a couple of notches above the usual airport hamburger? Check out Bay Bridge Airport (W29) in Stevensville, Maryland. It is located on the Eastern Shore of the Chesapeake about 90 nautical miles west of ACY and 20 nautical miles southeast of Baltimore. Bay Bridge has a 2,900-foot paved runway with pilot-controlled lighting and 24-hour self-service fueling using your VISA or MasterCard.

Less than a mile from the airport is the Kent Manor Inn (410-643-5757). Their friendly staff provide free shuttle service. Just pick up their direct-line phone outside the FBO. The food is good and the setting elegant but not snooty. A nonsmoking section is available. The Inn is open for lunch and dinner as well as for overnight stays.

AMENDMENTS

You can also get Kent Manor Inn to send over a car to pick you up at the airport. The Inn is more expensive than Hemingway's and is a bit more upscale. The food is great.

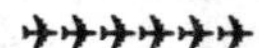

Kent Manor Inn is a little more formal than Hemingway's. The restaurant will provide transportation for the 1/2 mile ride to the Inn. A courtesy phone is located on the side of the FBO in the breezeway next to the pay phone. They have a great Sunday brunch.

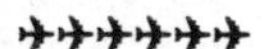

My wife and I flew to Bay Bridge Airport on 30 April '97 for her birthday. The airport is in fine shape, and the Kent Manor Inn courtesy phone brought a driver and van in 10 minutes. Located about a half-mile from the airport, it is also a nice walk.

The food was outstanding and not too expensive: appetizer, salad, main course, and dessert for two was $80.00. Reservations are a good idea.

The Kent Manor Inn is the perfect place to fly in for a romantic date. Outstanding food, great atmosphere and proximity to the airport make it a really special flying event. Call it the $100 filet mignon. The inn is also, well, an Inn, so for a truly memorable fly-in, plan to stay a night.

Fred's Grapevine 

If at W29 - Stevensville - Bay Bridge, and you're tired of the crabs, try Fred's Grapevine, about one mile east of the bridge. The great Italian food, reasonable prices, and congenial host are worth the walk. Call 410-643-4640 and tell him you're a pilot. He'll pick you up.

Westminister (Carroll County - W54)

Bullock's

Try Bullock's at Carroll County's Jack Poage Airport (W54). Located just north of Westminister, MD, Bullock's serves up some of the finest home-style cooking in the state. Their buffet dining is extremely popular, or you can order from the menu. There are no reservations. The building has been expanded over the last few years to accommodate its growing popularity. Try a Sunday morning flight to W54 in order to sample Bullock's breakfast buffet. Everyone should enjoy the enormous variety of food. Eggs, pancakes and creamed-chip-beef make a typical breakfast fare. You may keep it light with a variety of fruits and juices.

Stop into the FBO at Carroll County and say HI! to June and Amma before you take the short walk to the restaurant.

AMENDMENTS

Bullock's Airport Inn, 410-857-4417, is owned by Nancy and Don Bullock. Park near airport operations in front of the FBO, walk through the gate and head down the road. The restaurant will be on the left. Open seven days from about 7:00 AM to 8:00 PM. Good breakfast, lunch, and dinner fare at very reasonable prices. A lunch buffet is available Monday through Friday from 11:00 AM - 4:00 PM. A dinner buffet is available Monday - Thursday from 4:00 PM - 8:00 PM. Buffets are available virtually all day Saturday and Sunday. The prices range from $5.00 to $9.00 depending on day of the week and time of day. The standard menu items are typical of most restaurants with entree prices ranging from $1.00 to $5.00 for breakfast, $2.00 to $7.00 for lunch, and $6.00 to $11.00 for dinner. This place is well worth the trip for any meal of the day, any day of the week. Take the time to sign the Pilot's Guest Book on the way out.

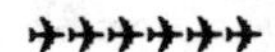

The restaurant, Bullock's, was very good. A great trip! Sorry we had not made the 30-minute ride from HGR sooner.

Bedford (Hanscom Field - BED)

A.P. Pizza 

The restaurant is a short walk from the east ramp in bldg. 1534. They have great pizza, calzone, subs, spaghetti, salads, and soft drinks for a very reasonable price.

The hours of operation are Mon. - Fri. 10:30 AM to 10:00 PM, Sat. Noon to 9:00 PM, Sun. Noon to 7:00 PM.

Beverly (Beverly Muni. - BVY)

Steffy's Diner 

There is a great diner about 1-2 miles from Beverly Airport's eastside ramp, Steffy's Diner. They serve a fantastic and really cheap breakfast. Follow the service road out of the east ramp, turn right and drive straight past three sets of lights, 1-2 miles total. Cross the RR tracks. Steffy's is on the right.

The East Side Cafe 

The East Side Cafe is now open! The food is excellent. Old WWII and later memorabilia is scattered throughout the restaurant. It is brightly lighted, with a west-facing view toward the ramp and runways. The transient ramp is located nearby.

AMENDMENTS

The East Side Cafe at Beverly Airport is in the process of changing hands, so check before you come in.

✈✈✈✈✈✈✈

It appears that the East Side Cafe is once again open at the Beverly Airport — under new management. It is open on weekends.

Chatham (Chatham Muni. - 0B6)

Crosswind

Chatham, on Cape Cod, has a great restaurant. The Crosswind operates only in the summer and is on the field. Provincetown, is a cab ride to many great restaurants.

Edgartown (Katama - 1B2)

Mel's Diner

Mel's is a typical airstrip diner located at the Katama Airport (1B2) in Edgartown, Massachusetts, on the island of Martha's Vineyard. The runways are a grass surface. The main runway of three is 3/21. It is 4,500 feet.

The food at Mel's is home-style. The setting, however, is spectacular, especially when experienced on the south-facing verandah, which overlooks the main runway. The field is part of a municipally owned preserve and is home to rare prairie grass and animal life. The very famous south beach is just beyond the dunes, where the airport provides parking for visiting planes at a nominal fee.

Fitchburg (Fitchburg Muni. - FIT)

Gene Collette's

Another local fly-in with a basic menu highlighted with great homemade pies for dessert. A good view of the active runway is available from the restaurant.

AMENDMENT

I find the restaurant at Fitchburg (FIT) to be one of the best family-style places in the area, second only to PYM. The breakfasts on the weekends are terrific. Usually they have specials on lunch prices and you can watch the plane traffic.

Hopedale (Hopedale Industrial Airpark - 1B6)

Airport Cafe

I went to the restaurant at Hopedale-Draper. It is on the runway next to the terminal ramp. They serve excellent, yet inexpensive lunch and dinner. In the same building is a nice family billiard hall. Bring your sticks.

Hyannis (Barnstable Muni. - HYA)

Mildred's Chowder House

Taxi to Hyannis Air Service and park. Walk down the access road to the main highway, turn right and walk about one block down for a total of about 3/4 mile. Great seafood served at a fine restaurant; look to spend $10.00 - $15.00 for a NICE dinner, or $6.00 - $12.00 for lunch.

Check the winds carefully. They get some real wicked winds in there, not too bad if they are right down the runway. I've taken off with winds 20 and gusting to 30, but it was straight down the runway.

The Cafe at the Airport 

The Cafe at the Airport is a pleasant, warm and cozy breakfast, luncheon and takeout restaurant that accommodates travelers, pilots and airport personnel. It is open 6:00 AM - 6:00 PM seven days a week and presents what can be described as a "wild blue yonder menu." It is appropriately decorated with artwork of aircraft and cape scenes.

All pilots are directed to the east ramp. The restaurant is in the main terminal building on the south ramp. It is accessible by a courtesy van provided by Airport Operations on 122.95.

The menu provides a variety of specialty sandwiches and burgers cooked to your specifications. The dessert specials are always a delight. All items are appropriately named after local aviation businesses. The Cafe service is fast, friendly and courteous for the busy airport environment. Pricing reflects the service and quality of the food selection.

Lawrence (Lawrence Muni. - LWM)

China Blossom 

Try the China Blossom across the street from LWM. It has a great buffet. It is not exactly a hamburger joint, but it is a good place to eat and it is within easy walking distance from the airport. One other update; the restaurant at the airport itself has gone by the wayside.

Joe's Landing 

Finally a new restaurant, Joe's Landing, has opened at LWM (Lawrence, MA). It has been open for two weeks and replaces the old restaurant that closed many months ago. It is a Greek-oriented sandwich place and was not too bad when I tried it yesterday. It is located in the terminal building.

LWM has two long hard surface runways, tower control and lots of parking. It is also my home field, where the Astrocoupe lives, so it is nice to have a place to eat. Now if only I could get a new windshield on the Astrocoupe, I could eat somewhere other than LWM!

Treadwell's Ice Cream 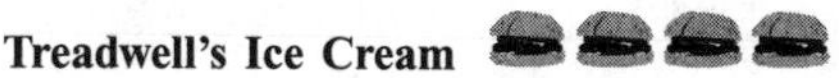

Anybody flying into Lawrence Airport has got to walk over to Treadwell's Ice Cream, right at the end of the runway. The only year-round ice cream stand I know of where you even have to wait in line in January.

Martha's Vineyard (Martha's Vineyard - MVY)

Airport Cafe 

Martha's Vineyard (MVY) has a nice restaurant. It is not as busy as Nantucket and has good food and nice people.

Nantucket (Nantucket - ACK)

Airport Cafe

Nantucket is "simply the best." It has good food, a nice staff and, of course, lots of the tourists coming though.

New Bedford (EWB)

Wings

There's a nice restaurant located in the terminal building at New Bedford (EWB). This is my "home" field so I may be a little biased. Food and beverages are very good. Outdoor seating is available during warm months. New Bedford is also a nice stopover if you're on your way to Martha's Vineyard (MVY) or Nantucket (ACK) Islands.

Orange (ORE)

Coffee Shop

Try the coffee shop directly across the street. It's good and only a 300-yard walk from the tie-downs.

Plymouth (Plymouth Muni. - PYM)

Plane Jane's Cafe

This is a friendly little fly-in supported by a well-maintained 3,500-foot runway. It's located in the second story of the FBO building, with an outdoor deck that is great for watching the comings and goings of fellow pilots. The food is basic fare, but the atmosphere of this local fly-in is great, particularly on a warm day when the deck can be enjoyed.

AMENDMENT

The description of the restaurant at Plymouth Municipal Airport, Plymouth, MA, is right on. It's a great place to eat, and the deck provides an excellent view of the flightline. My only addition is that the restaurant is now called Plane Jane's Cafe.

Southbridge (Southbridge Muni. - 3B0)

Airport Cafe

Hartford's EAA Chapter 166 held a breakfast fly-in at Southbridge Airport last Saturday. There is a '50s style diner located on the field with an outstanding breakfast menu at very reasonable prices. Haven't tried their hamburgers but if their breakfast is any indication, they must be great. Portions guaranteed to put you "over gross" and good service to boot. An outside eating area is available.

Stow (Minute Man Airfield - 6B6)

The Air Field Cafe

The Air Field Cafe is a rare gem in airport restaurants. The food is "homestyle," provided your mother was a great cook. The chefs use local products whenever possible and fresh ingredients. On most days the locals outnumber the pilots, a sign of a good airport restaurant.

The Cafe is open from 7:00 AM to 3:00 PM on weekdays and 7:00 AM to 4:00 PM on weekends. It is closed on Tuesdays. Breakfast is served until 11:30 AM each day. The lunch menu includes salads, sandwiches, homemade soups and stews. There are always daily specials listed on the blackboard. Be sure to leave room for dessert.

Minute Man Airfield (6B6) is northwest of Boston. The runway, 3-21, is 2,743 feet and paved. There is also a 1,600-foot gravel strip used by ultralights and STOL planes. Fuel is available 10:00 AM - 4:00 PM daily. Unicom is on 122.8. The cafe phone number is 508-897-3934. The airport number is 508-897-3933.

Westfield (Barnes Muni. - BAF)

The Flight Deck

Two long runways, interesting National Guard traffic, currently A-10s, and an excellent classic home-cooking breakfast and lunch restaurant on the field make this a great landing. Soups and burgers are recommended. The atmosphere is very pleasant.

AMENDMENTS

I would most certainly recommend The Flight Deck, located at Barnes Municipal Airfield in Westfield, MA, for a pleasant stopover to enjoy a bit of lunch. I'm never quite early enough to try their breakfast! Be sure to ask about the soup of and the special of the day as they are often worth a try. If you truly desire to taste the wonder of what a "$100 hamburger" can be like, order the "A-10 and a Coke." Be sure, however, that your plane is not close to its weight limit. Before you depart this fine eatery, be sure to say hello to Effie and crew and be sure to come back again.

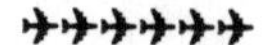

I stopped at BAF last week. They are now under new ownership and are open very early, before 7:00 AM, for those who want to indulge in their tremendous home fries, pancake delights, etc. I sampled the daily special, a seafood roll and 15 bean soup. Both were delicious, and $5.00 covered the tab. I stopped counting the soup beans at 11, as a sprig of parsley got in the way. Parking at the terminal building is very convenient. The controllers at BAF are first rate.

I landed between departures of two abreast A-10s. Good practice at wake turbulence avoidance. Generally, BAF is hard to beat!

Chesaning (Nixon Memorial - 50G)

Burger King

The Burger King and an independent ice cream stand are within ½-mile walk.

Coldwater (Branch County - D96)

Airport Inn

One of our very favorite places is D96, which is in Coldwater. One day the waitress even drove us back to our plane because it was snowing so hard we couldn't see our wingtips from the runway! Glad we were IFR rated.

Flushing (Daltons - 3DA)

Airport Cafe

Park on the grass at the north end of runway 36. The Cafe is a good family-dining establishment.

Gladwin (Gladwin - GDW)

The Peppermill Restaurant

A good place to go for a reasonable breakfast is The Peppermill Restaurant in Gladwin (GDW). Their varied menu of eggs, omelets and pancakes has always left us with a loss of appetite. It's a typical small town restaurant. It is crowded after church on Sunday. The airport has a courtesy car although it has seen better days! It is only a ½-mile walk if the car is not available.

Jackson (Jackson County - JXN)

Don's Airport Restaurant 

Don's Airport Restaurant is located on the Jackson County Airport (JXN). He serves a good breakfast and has daily lunch specials.

Kalamazoo (Battle Creek - AZO)

Bravo's 

Land at AZO and go to an FBO called Kal-Aero. Across the street from the airport is a very good restaurant called Bravo's. It has excellent Italian food and is a big step up from the usual airport dining. It is NOT open for lunch on Saturdays but is open all other times.

Ludington (LDM)

Gibb's Country House 

I flew into LDM from the Detroit area just to visit Gibb's restaurant. The nice people at the FBO called the restaurant for us. As expected, one of the managers of this family-owned-and-operated establishment promptly came out and picked us up in their minivan. Just as it was on our last visit, the experience was worth the trip. The quality of the food, the atmosphere, the decor, the friendly staff, and the conversation with the family members all show their commitment to quality. Full menu or buffet, the food is homemade, right down to the apple dumplings. You won't have room, but you have to eat one anyway. The sticky buns are a local legend — really!

Macinac Island (Macinac Island - Y84)

Grand Hotel 

This is a MUST trip on a midsummer night. A horse-drawn taxi will bring you to the hotel from the airport. Suit coats are required for dinner. The ambience is grand dining, complete with a dance orchestra.

For night departures, be prepared for immediate flight by instruments, as there will be no horizon with liftoff and flight over water.

Marshall (Brooks Field - 5D8)

Schuler's Restaurant

This is a real delight! Located two miles north of Brooks Field in Marshall Just use the phone (616-781-0600) at the friendly FBO and a courtesy van will pick you up on the spot! Real dining pleasure is available from a full-featured menu at this world renowned restaurant. After dining take a leisurely stroll through Fountain Park; it's just across the street. For a glimpse of real Victorian architecture, take the Historic Home Tour. You won't be disappointed.

Mecosta (Mecosta Morton - 27C)

Country Lake Inn

The Country Lake Inn is just across the street from the tie-downs of this short, 2,010-foot turf strip. The food is good, featuring Big hamburgers and omelets. It is well worth checking out. The prices are better than fair. I recommend it.

Mount Pleasant (Mt. Pleasant Muni. - MOP)

The Embers

The Embers, "Home of the One Pound Pork Chop," is located about a mile from the airport at 1217 S. Mission St. On a hot day, call a cab (616-772-9441). It's definitely worth the trip!

Muskegon (Muskegon County - MKG)

The Brownstone Restaurant

Every time I go there, I ask myself, "What is this restaurant doing in an airport?" I'm referring to the Brownstone Restaurant in the Muskegon (MKG) NEW Airport terminal. It is unbelievable! GREAT food, reasonably priced! Unique sandwiches and soups, all made from scratch on-site. Unusual sauce and vegetable combinations that are out of this world. All those Oshkosh-bound travelers from the east coast should stop in here before crossing the big pond to the land of cheese. The tower is in operation during most hours. Airport radar services are available (old TRSA).

AMENDMENT

I visited the Brownstone Restaurant at Muskegeon Airport (MKG) and found it lived up to the extremely favorable review. We were quite surprised that this quality of restaurant would be located at a small commercial airport! The food was outstanding and the service exceptional. A kids' menu was available, with most items $1.25. I highly recommend the restaurant to anyone!

Napoleon (Napoleon - 3NP)

Napoleon Cafe

The Napoleon Airport is located SE of the city of Jackson. The Napoleon Cafe is about one block east of the field and can be counted on for good food and great homemade pies.

Owosso (Owosso Community - 5D3)

The Pines

Taxi to the end of runway 6, The Pines restaurant's aircraft parking area. Enjoy the buffet or order from the menu. Runway 6-24 may be closed in the winter and early spring, but transportation is available.

JR's

The Pines is no longer serving dinner. I found out last night. There is a good home-cooking spot about one block to the east of the Pines called JR's.

Pellston (Pellston Regional - PLN)

Brass Rail

Fly into Pellston and visit the Brass Rail for great food. On Fridays they have all-you-can-eat fish dinners. Check it out!

Plainwell (Plainwell Muni. - 61D)

Airport Cafe

Plainwell has a nice little restaurant on the field. The food is good, the people are nice and they are open 6:00 AM to 6:00 PM daily.

Pontiac (Oakland-Pontiac - PTK)

Cafe Max

Located next to the terminal ramp at Oakland Pontiac Airport (PTK), which is 10 miles west of the city of Pontiac. Great everything. I thought it was a little pricy at first, but you get your dollars worth and much more.

AMENDMENT

I stopped in for lunch after a recent BFR. Tasty burgers so large a weight and balance was required afterwards. I thought the prices were very reasonable and the service good. They have a large deck outdoors where you can eat and watch the bizjets. Definitely worth the four-burger rating!

Saginaw (Tri. City Intl. - MBS)

The Skyroom

Located in the terminal building on the second floor, the Skyroom Restaurant is eye level with 727s flown by United and Northwest. This is a full-service dining experience.

St. Ignace (Mackinac Cnty. - 83D)

North Bay Inn

The North Bay Inn offers family dining, buffet-style or off the menu. It is located on the SE corner of the airport. Aircraft parking is on the slope just off the runway.

Traverse City (Cherry Capital - TVC)

Trilium Restaurant

Though it is not located on the field, the Grand Traverse resort will dispatch a car to meet you at the General Aviation Terminal. Call a few hours in advance 616-938-2100, to make arrangements. After landing, ask the FBO to call Grand Traverse and advise them of your arrival. By the time you're tied down and have your fuel order placed, the van should be there.

The Trilium Restaurant is on the top floor and has a terrific view of Grand Traverse Bay. We recommend going on a night the full moon rises about sunset. A table at the west window will give you the treat of a spectacular moonrise while the sun sets.

Traverse City (Sugar Loaf Village - Y04)

Sugar Loaf Village Resort

This is primarily a ski resort but it is great in the summer, too. The ski hill is immediately adjacent to the airport. The Sugar Loaf Resort is on the airport so there is no problem getting there. The restaurant is full service. If you like, stay overnight at the hotel 616-228-5461, or rent a condo. This is a truly beautiful resort.

Alexandria (AXN)

Arrowwood 

Alexandria (AXN) is located in the middle eastern part of Minnesota. It is the jumping off spot for much of the Minnesota lake country.

Arrowwood, a Radisson Resort, has 450 acres on Lake Darling. It offers horseback riding, indoor and outdoor tennis, golf, marina, indoor and outdoor swimming pools, sauna, cross-country skiing, snowmobiling, skating, and over 15,000 square feet of meeting space. Just four miles from the airport, they have a beautiful dinning room and deck, which looks out over the lake. They will pick you up and drop you off at the airport. Arrangements can be made by calling 320-762-1124.

Alexandria is also the manufacturing site for Bellanca Aircraft.

Appleton (Appleton Muni. - AQP)

Shooters Bar & Grill 

Appleton is home to Shooters Bar & Grill. It is a reasonable distance from the tie downs, only one mile west of the airport. They serve a good burger basket and more. Call ahead to verify availability of transportation (320-289-1100) if you're not up for the hike!

Backus (Backus Muni. - 7Y3)

The Corner Store 

The Corner Store has good food and is right across the highway from the airport.

Bemidji (Bemidji-Beltrami County - BJI)

Holiday Inn

On the way up to Canada from the Twin Cities this last summer my friend and I, both pilots, stopped for fuel, for both ourselves and the plane, in Bemidji, MN. It is a large airport with both a GPS/NDB and an ILS approach. There is one large FBO on the field. To our delight, there was a hotel about 1-2 blocks away from the airport. In the hotel there is a restaurant. The name slips my mind, but it had good food. They have a full menu, ranging from shrimp cocktail for starters to NY strip steak for your main course. You can also get an array of different burgers. It's a little more pricey than most places to fly into to eat. You can expect to pay between $10.00 - $20.00 per person, but it's good food and it's only a short walk from the airport.

Stats Sports Bar and Grill

I believe the hotel that Chris Malo referred to in his review of the airport restaurant at Bemidji is the Holiday Inn.

Another good restaurant in the Bemidji area is Stats Sports Bar and Grill, located in downtown Bemidji. It is a good distance, approximately five miles into town, but the FBO there always seems to have a car available if you put gas in it. Stats has really, really good food, and the service is always good. The prices are extremely reasonable, from $6.00 to $9.00 for most meals. Down below the restaurant is a gift and sweet shop. The adjacent bookstore makes for a nice little trip.

The FBO also has a pilot shop with some nice things in it. They are friendly and helpful.

Brainerd (Crow Wing County - BRD)

The Airport Cafe

The Airport Cafe has good food and is right on the airport!

Duluth (Duluth Intl. - DLH)

Airport Restaurant

The FBO gives rides to the terminal building for fancy grub at good prices. It is a nice place for dinner.

Duluth (Sky Harbor - DYT)

Grandma's Saloon and Deli

Sky Harbor is located on a narrow sand peninsula in the southwest end of Lake Superior, where the St. Louis River meets the lake. This destination is beautiful and relaxing in the warmer months of summer. The airport features both a paved runway and a seaplane area in the bay. Either way, taxi to the dock or FBO, walk behind the FBO and over the sand dune, and there before you stretch the beautiful lake and miles of clean sand. Though the water remains a chilly 50 degrees even in summer, tourists with hearty souls and some of the locals venture in on hot days. The airport is about a three-mile cab ride from the Duluth harbor, which is a seaport for oceangoing vessels and features shopping and restaurants, including Grandma's Saloon and Deli. A couple more miles across the lift bridge and you're in downtown. From Duluth north, the landscape rises from the lake and is very scenic. A great day trip or weekend visit!

Eden Prairie (Flying Cloud Airport - FCM)

The Lion's Tap

Just five minutes south of Flying Cloud Airport in Eden Prairie, Minnesota, is a wonderful burger place called The Lion's Tap. You will experience great service, a fun atmosphere, and most important, excellent burgers!! They have a limited menu, but they make one heck of a burger! Please take the time to drop into Flying Cloud (FCM) and ask anyone there how to get to Lion's Tap. Trust me, they'll know!

Faribault (Faribault Muni. - FBL)

The Lavender Inn

The Lavender Inn in Faribault is an excellent place to relax after a long or short flight. It's a very comfortable atmosphere along with your choice of savory food prepared to your pallet's requirements. The restaurant is located 1/2 mile east of the airport.

Luverne (Aaneson Field - D19)

Magnolia Steak House and Lounge

If you're willing to walk a mile, you can enjoy a great steak or burger at the Magnolia Steak House and Lounge. The airport has a courtesy car if you'd rather not walk.

Scotty's

Scotty's is a bar and grill located very near the D19 airport. It is owned by the airport operator, so pilots are very welcome and will enjoy this aviation-friendly atmosphere.

Maple Lake (Maple Lake Muni. - Y33)

Maple Lake Cafe

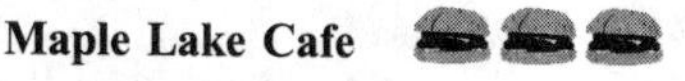

Maple Lake Cafe is a ¾-mile walk from the airport. It offers good food and a nice hometown setting. Prices are very reasonable. A $5.00 dinner special is always available.

Olivia (Olivia Muni. - Y39)

Sheep Shedde

The Sheep Shedde is an old English restaurant about two blocks from the airport. Good lunch buffet on weekdays, Sunday brunch from 10:00 AM to 2:00 PM. You may walk or call 320-523-5000 for a ride.

Princeton (Princeton Muni. - PNM)

Pine Loft Restaurant

The Pine Loft Restaurant is about 1/4 mile from the north end of the old runway. It opens at 11:00 AM every day.

Redwing (Redwing Muni. - RGK)

St. James Hotel

Fly into Redwing Airport, designation (RGK). Redwing provides a courtesy car, or the St. James will pick you up. It's 4-5 miles away and worth the ride. The St. James, which was built in the 1880s, overlooks the Mississippi River. Their Sunday brunch is a treat!

St. Paul (St. Paul Downtown - STP)

Northwind's Restaurant

Northwind's Restaurant has excellent sandwiches, burgers, salads, and a surprisingly great New England clam chowder served with soda bread. The restaurant is right below the tower in the administration building; you can park your airplane and walk right in. Sit by the window, eat lunch, and watch the airplanes go by. They are open from 8:00 AM to 3:00 PM daily.

Brookhaven (1R7)

The Fish Fry

On a recent cross-country trip from Texas to Virginia and back again, we stopped at Brookhaven, MS (1R7), both going up and coming back. If I was ever in doubt about the meaning of Southern hospitality, I am not anymore. On the way to Virginia we just stopped for gas at Brookhaven. The FBO, Al Jordan, directed us to the gas pump, filled up our plane for us, then treated us to soft drinks and snacks. The airport terminal building and ramp were extremely well kept; everything was spit polished. On the way back from Virginia we stopped for lunch and fuel. Al Jordan handed us the keys to the courtesy car and directed us to a place not far from the airport called The Fish Fry.

The Fish Fry, as the name implies, fries everything, including DILL PICKLES! I had never had fried dill pickles before, but as it turned out they were very good and went well with the fried catfish and Cajun fried potatoes. They did not fry the cole slaw! The Fish Fry is a plain, small town restaurant that serves good food for a reasonable price. Lunch was about $5.00 per person.

Corinth (Roscoe Turner - CRX)

Margel's

I landed at Roscoe Turner just at sunset. This is one of the best GA airports I have ever encountered. It is ultramodern, well-kept and uncrowded. While Southern hospitality wasn't born here, it is re-invigorated. I told the lineman that I was staying the night. Before I got to the desk, they had a hotel room reserved for me at the Comfort Inn, the rate a budget-pleasing $36.00! How would I get there? Easy! They handed me the keys to one of their airport cars, a reasonably new Ford AeroStar van. Where's the best place to eat? "Why, that would be Margel's."

In the middle of a restored section of downtown is Margel's. In New York or San Francisco you'd have to call it a CAFE. It is TRENDY! The food is superb! I started with a satisfying potato soup and had an onion-glazed roasted chicken as an entree. It was all delicious. The bill, including tip, was less than $10.00. I'll go back!

Gulfport (Gulfport-Biloxi Regional GPT)

Montana's BBQ

The FBO provided transportation for the three-minute trip from the ramp. The food is all-you-can-eat buffet style — the BEST BBQ you can imagine for only $6.95 before 3:00 PM and $8.95 after. The FBO requires a fuel purchase to use the courtesy vehicle.

Hattiesburg (Pine Belt Reg. - PIB)

Approach Zone

Try out the tasty treats at the Approach Zone at the Pine Belt Airport in southern Mississippi. It is just north of Hattiesburg. They're open from about 10:00 AM to 5:00 PM. Good burgers, soup, and sandwiches are available. My personal favorite is the "po-boy" sandwich.

McComb (MCB)

The Dinner Bell

It's a little tough to find but I'm sure if you call the Dinner Bell in McComb, they would be more than happy to come pick you up. There is also a courtesy car on the field. Everything is served on a lazy Susan around a huge table. Great food at a low price!

Natchez (Adams County - HEZ)

The Lady Luck Casino

The Lady Luck Casino, a riverboat on the Mississippi River, is a great place to go. Land at HEZ and the FBO will call the casino to have you picked up FREE of charge. They have a large buffet and lots of gaming. With a little luck you can go home with full stomachs AND full pockets.

Ava (AOV)

Rudy's Village Inn Restaurant 

Rudy's offers standard smalltown blueplate cuisine, with a running special of ribs or fish buffet. I think the ribs were finished off in a crockpot. The meat fell off the bones, and the bones were almost tender enough to eat. My daughter had a cheeseburger with fries, true pilot in training.

AOV is in Ava, Missouri, east of Springfield and nine miles east of the Dogwood VOR. The hike to Rudy's is 1.1 mile along a smooth paved backroad, with a pretty scenic pastoral view along the way. Get there during daylight or call ahead, 417-683-3765, and the airport manager, Bud Miller, will probably drive you over if you really want to miss the pleasant Ozark air. The phone number for Rudy's is 417-683-5465. If you need anything else, Wal Mart is next to Rudy's.

Cape Girardeau (Cape Girardeau Muni. - CGI)

Cape Airport Restaurant 

This is a nice restaurant in the terminal building serving a smorgasbord at lunch. The food is good and the prices are reasonable.

Carthage (Myers Muni. - H20)

A Steak House 

There is a franchise steakhouse, something like a Golden Corral, just a block away from the Carthage Airport. The food is just OK. I took my wife there on our first date. For what it's worth, there is a WalMart Super Center right next to the runway.

Cuba (UBX)

Porky's Sports Grill

Porky's Sports Grill is located approximately one mile east of Cuba Airport. The food is very reasonably priced, and they have the best hot wings around (Hot!!). The food ranges from pizza and pasta to steaks, sandwiches and even seafood. You may want to call them for info on picking you up at the airport. (573-885-2749)

Dexter (Dexter Muni. - DXE)

Airways Restaurant

The restaurant on the north end of the field is open for breakfast and lunch. I paid $3.85 for a generous country fried steak lunch. Dexter is just west of the Mississippi in southeast Missouri. I plan to make it a regular stop.

The Hickory Log

I made an unscheduled stop at Dexter and caught a ride into town with the owner of The Hickory Log. I got the impression that regular runs are made to the airport to pick up fly-ins. I had to leave my airplane there, and a local gentleman gave me a ride to Cape Girardeau, which is a loooong way off. A nicer, more helpful bunch of people would be hard to come by.

AMENDMENT

The Hickory Log is a regional favorite offering the most excellent ribs! There used to be a courtesy car at the airport.

Garvois Mills (Habour - 7L7)

Resort Restaurant

Now this is one place I will never forget, nestled on the corner of the Lake of the Ozarks. A short walk down the hill from the runway brings you to this small resort restaurant where you can get Sunday brunch for a ridiculously small amount of money. Small runway, with a big grade sloping down toward the lake if I remember correctly. When you blast off after gaining 10 pounds at brunch, you'll like that downhill grade.

AMENDMENT

A friend and I flew to Harbour yesterday to try one of those $100 hamburgers with a "four-burger" rating. During our flair we see this sign, $7.50 LANDING FEE. After walking to the restaurant, we see Closed. Found out, as we should have guessed, it's only open seasonally. It won't open until the 1st week in April. We were too hungry to wait that long. Pilots should be aware of the landing fee and that it isn't open all year. It does look like a nice place.

Lincoln (Lincoln Muni. - MO11)

Papa Joe's

A gorgeous grass strip serves this tiny Midwest town of Lincoln. This place will take you back to the days of taildraggers. A short walk will bring you to a bunch of diners and a Pizza Hut. I've flown here quite a few times in a 150, and I must say, I love this little 'port!

AMENDMENT

Landing at Lincoln's grass strip is worth the trip alone, but the short walk to Papa Joe's eatery for breakfast or dinner is a real treat. Great food, very friendly folks and no calorie counting allowed!

Malden (Malden Muni. - MAW)

Airport Cafe

This 1950s-type diner is on the field. The place is long on atmosphere, black and white checked floor and all. Open daily from 11:00 AM - 2:00 PM, it offers Karaoke on Friday nights. They call it "Fly-in Karaoke."

Osage Beach (Linn Memorial - K15)

The Kenilworth House

At the Lake of the Ozarks there are a great many outstanding restaurants. One of the finest is the Kenilworth House, which is located off the end of RWY 32 in Osage Beach. The restaurant is only a couple hundred feet from the airport terminal. The food is great and reasonably priced. The prime rib and pork tenderloin are excellent and the fresh baked breads and desserts are wonderful.

For those wishing to stay a little longer, there is a Comfort Inn within easy walking distance. The FBO has rental cars available. For more information contact the Kenilworth House at 573-348-5959 or Grand Glaize Airport-Airlake Aviation at 573-348-4469.

Lil' Rizzo's

We fly into K15 several times each month to enjoy the Lake of the Ozarks. Lil' Rizzo's always has the best pizza and pasta in Missouri! If you're in the mood for courteous service and a great meal, drop into Lil' Rizzo's locations next time you're at the Ozarks.

Lil' Rizzo's is within walking distance of the Airlake terminal building. If you need directions, call ahead: Lil' Rizzo's Restaurants, 573-348-0304. See 'ya there!

St. Charles (Smart Field - 3SZ)

Mary's

One of my favorite burger joints on an airport in eastern Missouri is Mary's at St. Charles County Smart Field (3SZ). The Missouri Wing of the Confederate Air Force considers Mary's their home kitchen while they're working on the maintenance and restoration of their WWII airplanes.

You can get your burger any way you like it, along with Mary's homemade soup, chili and pies. It gets pretty crowded around lunch time on the weekends, but there are always plenty of flyers to talk with and people to meet while you wait for a table to open.

AMENDMENT

St. Charles, MO, is about 20 nautical miles west-northwest of STL. The restaurant is called Mary's. Mary owns the place and does the cooking. It was originally in an A-frame building, but that was destroyed by the flood of '93.

They have great food and are located at about the center of RWY 18/36 along the west side of the ramp. All the food here is fresh and the burgers are very good. I would suggest the chili-burger for those with the adventurous spirit.

St. Joseph (Rosecrans Memorial - STJ)

Airport Cafe

The restaurant is right on the field. Their prices are really low, about $2.75 for burger, chips and pop, $5.25 for Salisbury steak, salad, mashed potatoes and gravy, and green beans. They sell altimeter clocks for $20.00. They were open on Saturday at 1:00 PM.

St. Louis (Spirit of St. Louis - SUS)

Blaney's

Blaney's Bar and Grill, on the field just north of the old tower, is easy to find. Just ask for progressive taxi instructions from the tower. There is a pilot supply shop across the street from the restaurant called The Outer Marker. It has a good selection of aviation books, test guides and other pilot stuff. Makes for a very leisurely stroll to burn off the lunch calories.

Sikeston (Sikeston Memorial - SIK)

Lambert's

My favorite fly-in and chow-down place is at Sikeston, MO (SIK), down in the bootheel of Missouri. The restaurant is Lambert's, home of the "throwed rolls". Just ask for the Lambert's shuttle when you are five minutes out; they'll be there when you arrive. They really do throw their giant hot rolls to you! You may find that you are out of the weight and balance envelope when you leave.

AMENDMENTS

Lambert's cafe has got to be one of the best restaurants I've eaten in. They are famous worldwide for their "throwed rolls". They are located in Sikeston, MO, about a mile from the airport. They have a special van that will come and pick you up at the airport. Every time I go, there must be 20 to 25 people landing just to go to Lambert's. To give you an idea how big they are, last year they averaged 520 dozen rolls per day, for a grand total of 2,246,400 individual rolls. That's the distance between St. Louis, Missouri, and Memphis, Tennessee. This is all in one restaurant in Sikeston,

MO. You owe it to yourself to find out more about this place because it has excellent food and atmosphere at low prices. The number to the restaurant is 314-471-4261. Happy Eating!

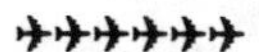

In addition to picking you up at the airport, the van driver takes pilots and their passengers into the restaurant through the kitchen, thereby avoiding the waiting line. You'll be seated at the first available table no matter how long the line is. When you are ready to leave, the driver takes you back to your plane and presents all in your party with a really nice insulated travel mug and a LARGE fresh sweet roll.

Springfield (Springfield Regional Airport - SGF)

Airport Restaurant 

If you like to fly into a busy airport, here's your chance. Activity can get very heavy. They have airline service, helicopter, and military operations going on. There is a restaurant in the main terminal, and a shuttle bus will bring you there from the GA parking area. The food is typical for a large airport restaurant.

Chico Hot Springs (Highway - Public)

Chico Hot Springs 

The best fly-in eating in Montana has to be at Chico Hot Springs. It is south of Livingston and just north of Yellowstone NP. Chico is a hot spring resort with a large outdoor hot pool, old style lodge. There are several other choices of accommodation, including standard hotel rooms and large cabins and chalets. The restaurant is well known as one of the best in the state. However; the best part has to be the airport, which consists of a roughly 1-mile-long stretch of road which dead ends at the lodge. As you approach, call Unicom and they'll send a truck out to close the road while you land. One of those things you always wondered about doing but never thought you could. It's a fun place with great food but don't expect to get out for $100!

Ennis (Sportsman's Lodge)

Sportsman's Lodge 

The Sportsman's Lodge is in downtown Ennis, Montana not at the nearby airport, 5U3 . Land behind the lodge, and taxi up to restaurant. Get permission to land and conditions first. I came in from the North over the gas station last time. It's right in the center of a very small town, near the main intersection. (406-682-4242)

The Continental Divide 

There's also a 5-star Continental restaurant named the Continental Divide in town. It has a superb reputation. You can walk a couple hundred yards to it from the strip.

Kalispel (City Airport - S27)

The Outlaw Inn

The Outlaw Inn is across the highway from the airport. It has an excellent restaurant. They provide a wonderful Sunday Brunch for about $10! Last time I flew over, they had salmon, red snapper, shrimp, roast beef, a great pork roast, some delicacy similar to a waffle, lots of pastries, perfect scrambled eggs, salads, fruits, apple crisp for dessert and a bunch of stuff I did not sample!

The brunch alone is good enough to justify a trip. Its adjacent location to Glacier National Park offers some of the most spectacular scenery I have seen between Arizona and Alaska!

Fred's Family Restaurant 

There are numerous motels, restaurants and a casino within a ¼-mile walk of Strand Aviation tie-downs at the east side of the field. Fred's Family Restaurant has great down-home food at very reasonable prices. The BBQ ribs are fantastic.

Three Forks (Three Forks - 9S5)

Custer's Last Stand 

Three Forks, Montana, offers a great little burger place, about 1/2 mile away from the airport, called Custer's Last Stand (406-285-9892). It is a wonderful place to get a burger and an ice-cold root beer in a mug. They also have a pretty good pizza.

The first week in August, the airport fills with activity as the Montana Antique Aircraft Association holds their annual fly-in. There are usually a dozen or so pre-WWII aircraft along with a hundred other private planes. On that Saturday morning, a good part of the group will usually fly, for breakfast, to some other nearby airstrip. On the field, for the weekend, a local restaurant sets up and serves breakfast, lunch, and one main course for dinner. When it gets dark, there is a hanger dance with a good C&W band. Hotels and motels are available in town. Plan on camping under your wing, to experience it all. There is not a better alarm clock than some radial-engined aircraft flying 30 feet over the top of your tent!

West Yellowstone (Yellowstone - WYS)

Mexican Cantina 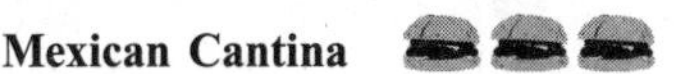

West Yellowstone, Montana, is the Gateway to Yellowstone National Park and happens to have a nice airport to service it. We found the Mexican Cantina, located in the terminal, to have some great Mexican food. Also located on the field is a campground just for flyers. If you want a fun excursion, fly in, camp, and then rent a car to explore the park. There is also a nice little fly-in held there in the summer time.

Alliance (Alliance Municipal - AIA)

The Elms

In Alliance, Nebraska, we ate at The Elms. It's a homey, Midwestern cafe. The service was friendly, the hamburgers tasty, and the pies looked tantalizing, but we would have had to do another weight and balance if we had eaten any!

While you're there, check out Carhenge, a unique car sculpture! The FBO at AIA provides a courtesy car.

Hastings (Hastings Municipal - HSI)

The Village Inn Restaurant 

The Village Inn Restaurant is only 100 yards away from Hastings (HSI) Municipal Airport. After your meal, enjoy the Leid Imax Theater. A courtesy car is available at the airport. This is an enjoyable community in the central part of Nebraska and an excellent airport with two runways.

Minden (Pioneer Village Field - 0V3)

Minden Country Club

A group of us were traveling to Oshkosh and ran across this airport by accident. After eating at the Minden Country Club, which is only ½ mile from the airport, we discovered The Pioneer Village Museum. The food at the country club was moderately priced and very, very good.

The Pioneer Village Museum claims to have a representation of every major technological advance in the past 100 years. I could not argue with them. It has about a hundred automobiles, lots of tractors, trucks, some aircraft, a steam train, merry go-round, and about everything else under the sun you can think of. Plan on staying a day, at least, to see the museum.

Minden is a nice friendly midwestern small town. Three courtesy cars are provided at the airport by the museum.

North Platte (Lee Bird Field - LBF)

Airport Cafe 

Lee Bird Field, North Platte, has a very fine restaurant on the field. Sundays can be crowded at noon.

Scottsbluff (William B. Heilig Fld. - BFF)

Skyport Cafe & Lounge 

I think I had the BLT, could have been a burger, but I was impressed with the little restaurant in the Scottsbluff, Nebraska, terminal. Good sandwich, good fries, good pie, good service.

AMENDMENT

I flew into Scottsbluff on an IFR training flight. The Skyport served the juiciest, tastiest hamburger I've ever eaten in a restaurant. Service was friendly. Nothing fancy but really good food. It is worth making the trip.

Sidney (Sidney Muni. - SNY)

Cabala's 

Request on Unicom (122.8) prior to landing that Cabala's be given a heads-up of your arrival. They will meet you at the airport with a van. Cabala's is an outdoor shop with a tiny sandwich shop in one corner. I suggest the hot smoked-buffalo sandwich. It is really good! If you are an outdoorsman, expect your lunch to include some equipment purchases. Their selection is hard to resist.

Wallace (Wallace Muni. - 64V)

Wallace Cafe 

Wallace has an excellent hometown restaurant located two blocks north of the airport (64V). Food, service, and atmosphere are enjoyable. The menu features breakfast, dinner, sandwiches, salads, baskets, and desserts. During the week they offer a lunch special. Hours are 7:00 AM until 8:00 PM Monday through Thursday. On Friday, they close at 6:30 PM and on Saturday at 4:00 PM. Naturally, they are closed all day Sunday.

York (York Muni. - JYR)

Chances "R" 

The York Municipal Airport (JYR) has an excellent place to eat. The restaurant is called Chances

"R" and is located three miles away. It was rated the No. 1 restaurant in Nebraska for 1995. If you have been there, you understand why. Order anything from a greasy hamburger to prime rib. On Saturday nights, they have a prime rib buffet. The York Airport has two courtesy cars for transportation. The two cars are usually being used every Friday and Saturday by pilots who fly in to eat. If both cars are out, Chances "R" will pick up at no charge anyone who wishes to fly in. Call 402-363-2660 to reserve one of the courtesy cars.

Carson City (Carson - O04)

Pookie's Final Approach 

My home base, Carson City, has its own restaurant on the field, a place called Hangar 17. I've only eaten there once, and the food was OK, nothing special, but not too bad.

AMENDMENT

Hangar 17 at Carson City Airport has changed hands. It's now Pookie's Final Approach. They're open for breakfast on weekends and lunch seven days a week. Generous portions and great food.

Elko (Elko Muni. - EKO)

Zappata's 

Zappata's is approximately ½ mile from the airport, an easy walk. They serve very good Mexican food.

Hawthrone (Hawthrone Muni. - HTH)

The El Capitan Casino 

Hawthorne, NV, is right at the south end of Walker Lake, which is noted for being the home of giant cutthroat trout. The El Capitan Casino has good food and gambling. They will send a courtesy van to the airport to pick you up. After lightening your wallet, they will even take you back.

Mesquite (67L)

The Players Island

Mesquite, NV (67L), is located on the NV/AZ/UT borders. Inside the airport building are telephones that are connected to three hotel/casinos. One of our favorites is The Players Island. They will send a courtesy vehicle to pick you up at the airport, usually a Lincoln or a Limo. They have breakfast, lunch, and dinner buffets that are very good. Mesquite also has a couple of golf courses.

North Las Vegas (North Las Vegas Airport - VGT)

The Good Ole USA

North Las Vegas Airport (VGT) has a brand new terminal. Its restaurant, The Good Ole USA, has an excellent second-story view of the ramp. More than just burgers and grill items, they've got the good stuff.

Reno (Cannon Intl. - RNO)

Amelia's

The restaurant I visit most often is Amelia's in Reno, Nevada. Fly in and park at the Reno Jet Center. Amelia's is right inside. The food is pretty good, and the specials are usually very good. As an added bonus, you may get to park next to a military jet. They fly in from time to time.

Wendover (Wendover Airport - ENV)

Peppermill Casino

This airport is right on the border of Utah and Nevada. Since Utah does not allow gambling and Nevada does, it attracts a lot of Utahns on the weekend. The Casino offers transportation to and from the airport. When I think about the $100 hamburger, this is where we go the most. Reach them at 702-664-2255.

AMENDMENT

The Nevada entry "Wendover (ENV)" is actually an airport in Utah, but I see your reason for putting it under Nevada. However, to be complete, you should at least add it to the Utah list as well.

Yerington (Yerington Muni. - O43)

The Hangar Cafe

The Hangar Cafe at Yerington Airport (O43) looks like your basic truck stop/greasy spoon, but the breakfasts are very good and the servings VERY generous. You may have to refigure weight and balance after eating a stack of pancakes.

Joe Dini's

Joe Dini, the Speaker of the Nevada State Assembly, operates a casino and restaurant in downtown Yerington, just a ¾-mile walk from the airport. Joe makes sure the food is fresh and the prices are right. It's usual American fare, but the freshness of the ingredients makes all the difference. Walk another ¼-mile south to visit the Lyon County Museum on the same side of main street before you leave. On the way home, fly over the huge Arimetco copper pit and leach pads. If you are coming from or heading north, fly true north 10 nm to the junction of US 95A and the Southern Pacific Railroad. Follow the latter to the northwest up to the Carson River canyon, the ruins at Fort Churchill, and over to Lake Lahontan before you leave the area. A scenic flight over historic country is well worth the trip. It's only 80 nautical miles from South Lake Tahoe or Reno.

Hampton (Hampton Field - 7B3)

Airport Cafe

Just over the Massachusetts border into New Hampshire, there's a small airport called Hampton Field (7B3). It's a small field, 2,100-foot grass strip, located about 3-4 miles from the beaches of Hampton, NH.

They just opened a restaurant there about two years ago that serves great burgers and sandwiches. Connected right to the flight shop office, this place makes a great hangout on weekends. It's a small and cozy airfield, just great for conversation.

Jaffrey (Silver Ranch - AFN)

Kimbell's

This small airport is located just a few miles from the Dillant-Hopkins Airport. Although there isn't a restaurant located on the airport, there is an excellent homemade ice cream shop within a two-minute walk. The serving sizes at Kimbell's are incredible, so don't buy a large anything unless you have a huge appetite or haven't eaten that day. There are several benches out front which provide a great view of departing aircraft. They are not open in the winter and I am not exactly sure when they open for the season. If you call the Jaffrey Airport, I'm sure they can provide you with that information.

AMENDMENT

Kimbell's ice cream stand reopened for the season about a month ago (late April) and is once again dishing out huge servings. There was no visible difference in size between the small and the peewee cones during a Memorial Day weekend visit with friends. Anything larger will certainly put you over gross. The field is interesting enough without that! They also have food service at the restaurant I had the house burger special, and it's just as ample as the ice cream servings.

AMENDMENT

Went to Kimbell's on 4/27/97. Had a lobster roll that was out of this world. It must have had 1/2 pound of fresh lobster inside, no fillers, pollock substitutes, or the like. Unusually served with cottage fries and watermelon, no less. Priced at $8.75 and worth every penny. All in all, a heck of a way to spend a lovely spring day!

Keene (Dilliant-Hopkins - EEN)

Campy's Country Kettle 

The approach to this well-maintained airport is highlighted with fantastic New England mountain views. It's nice any time of year, but in the fall, when the foliage is changing, it is particularly inviting. The lunch menu is very simple and includes an all-you-can-eat buffet. I'm guessing that they change the menu for dinner and attract neighborhood locals, not just fly-in customers. They have live entertainment on the weekends.

AMENDMENTS

Campy's Country Kettle has always been open 6:00 AM - 2:00 PM daily seven days a week. Campy's is located through the fence and up the hill at the end of runway 2 on the right.

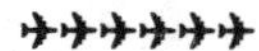

We were in Keene last weekend, July 5, and the Country Kettle is open again. It is just outside the fence behind the (closed) Emerson Aviation on the northeast corner of the airport. The food was OK, and the prices were really cheap. My hamburger with fries was quite good. My wife's egg-salad was overdone with mayo. The bill, including tip, was less than $10.00. The menu looked pretty good, with breakfast, lunch and dinner items, including a salad bar.

Laconia (Laconia Muni. - LCI)

Patrick's Pub

Call a taxi or call Patrick's Pub. They run a BIG limousine to the airport, upon request. Do check with Patrick's Pub before you leave, though. I found out the hard way that they no longer run the limousine on Monday nights. You can walk, but it's about 1 1/2 miles. Patrick's Pub is a great Irish restaurant, suitable for evening dining. I understand there are also dinner cruises available on the lake. I don't remember the place that sponsors this. I gotta' try it sometime.

AMENDMENT

Patrick's offers several excellent menu choices in the $10.00 – $20.00 range. If you want to walk from the field, I suggest you park the plane at the east end of the taxiway, well off to the side. Many folks do this. Look south and see Route 11 running along the runway heading. Follow it to the east, about one mile, good to stretch your legs, and you'll find Patrick's on your left under the glidepath for RWY 26.

Lebanon (Lebanon Muni. - LEB)

Tailwinds Restaurant 

The Tailwinds Restaurant in Lebanon, NH, is in the very small airport terminal. It provides low- to medium-priced cuisine of good quality. Try to get a window table to enjoy the nice view of the runways. The Tailwinds is closed on Sunday. Tell ground control (121.6) you want to go to the restaurant. They'll tell you where to park.

AMENDMENT

We also tried the Tailwinds restaurant at LEB and it is outstanding. A very different menu with things like "jalapeno poppers," buffalo wings, crab burgers, garden burgers, pizza burgers, seafood soup, and all kinds of salads. They offer a nice dinner menu with entrees like pasta pomodoro and smoked salmon penne, along with more traditional chicken, steak and seafood dishes. They are open from 11:00 AM to 11:00 PM daily, and CLOSED on SUNDAY. This is a great place to bring friends or a special guest for a nice evening out. If you fly in at night, watch out for the hills on the southeast sector of the airport.

Nashua (Boire Field - ASH)

The Crosswinds 

Located in the blue building at the Texaco sign. You can park out front. They serve standard fare at reasonable prices for breakfast and lunch daily. The Crosswinds is very busy on the weekends!

Atlantic City (Bader Field - AIY)

Atlantic City Boardwalk 

The Atlantic City Boardwalk is only one mile away. Taxis are readily available.

Blairstown (Blairstown - 1N7)

Runway Cafe 

The Runway Cafe is a smallish grill serving typical airport fare. They have limited hours of operation. Blairstown has a glider operation so be careful in the pattern.

AMENDMENTS

I feel that the review given to The Runway Cafe in Blairstown, NJ, is somewhat unfair. The hours are inconvenient at times, 6:00 AM to 6:00 PM, but the food is excellent. They exhibit a clear commitment to quality food at a reasonable price. Almost everything is homemade, from the soup of the day to incredible pies! They are also making an effort to offer healthy choices with the addition of vege-burgers to the menu as of late.

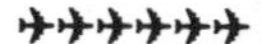

I agree with the comment that this restaurant is underrated. The service is friendly and the homemade pies are not to be missed. The only miss on a "five" is the limited seating capacity.

The Runway Cafe, 908-362-9170, is on the field. SRO on Sundays because so many locals turn up.

Caldwell (Essex County - CDW)

94th Bomb Squadron

It's Essex County on the sectionals, but the tower goes by the name of Caldwell and underlies NYC Class B airspace. This is one of a chain of restaurants that are almost always built on or adjacent to the airfield, usually with direct airport access. Views of the runway are typical. They have an excellent menu and tend to have moderate to higher prices. This chain is several steps above your typical $100 hamburger. The motif is a WWII bombed out English farm building with various aviation artifacts.

AMENDMENT

This 94th Bomb Squadron is one of 72 in the U.S. There are 42 company-owned and 30 leased units. Parking is at CAS, Caldwell Air Service, on the ramp immediately next to the restaurant. Walking is closer than from the auto parking lot. Landing fees at CDW are waived for SEL and light MEL aircraft.

NY Steakhouse & Pub

I've got the place for steak eaters, and to top it off, it couldn't be more accessible. It is directly across from the tower at Essex County Airport/Caldwell Air Service in Caldwell, NJ. The name is NY Steakhouse & Pub, and they serve three 24-oz. steaks for $10.95 each. They are quality steaks and priced right. When you go, try the coconut shrimp and/or BBQ ribs. They also have chicken, blackened or BBQed, salads, live lobster out of the tank, blackboard fish specials, and finger food, including chicken fingers, mozzarella sticks, buffalo wings to die for, and the other normal stuff. Burgers? Chili, bacon, bacon cheese, cheese, regular and mushroom for those interested in burgers only. The fries are seasoned just right, on the spicy side, that is, and the chicken pot pie is served in a pastry pot that is just as good as the steaming chicken center. The prices are ALL economically minded, soda refills are free, and happy hour is $1.00 drafts and half price mixed drinks, M-F 1700-1900.

It's about time someplace opened CLOSE to an airfield that is truly worth going to.

Cross Keys (Cross Keys - 17N)

Airport Cafe

The restaurant is located at the approach end of 27. The food is good, service friendly, and prices are reasonable. If you can get a window seat, you can rate everyone else's landings. Watch for parachuting! The landing zone is just to the north of the runway at midfield.

Lumberton (Flying W - N14)

Avion 

The Flying W (N14) Airport has a restaurant/bar/motel/FBO right on the airfield. The bar at the Avion restaurant has a nice jukebox, piano, and single pool table as well as an interesting but small local clientele. The restaurant has some Italian and Mexican dishes, but the cheeseburgers/cheesesteaks/fries are all yummy. The jalepeno poppers are tasty, too. It has a 50s Western kitsch

motif. Without drinks, just perhaps an appetizer/burger/Coca-Cola, expect $12.00 per person.
AMENDMENTS

Earlier today my wife and I returned from a short flight to Flying W. We met friends from south Jersey for Sunday brunch. I would recommend this highly for a Sunday breakfast trip. Brunch is served at 10:00 AM and costs $12.95 per person. The four of us enjoyed an extended, relaxed and very filling experience, which gave us lots of time to catch up and enough food to last into the early part of next week.

A friend and I stopped this past Sunday. The brunch is 10:00 AM to 2:30 PM, only on Sundays. The quality and quantity are great. I know of none better. I am almost afraid to drop this tip. I may now have to wait in line to get in. Reservations 609-267-8787.

Manville (Central Jersey Regional - 47N)

Pizza & Pasta 

Located just at the entrance to the airport, about a ¼-mile walk from the FBO is Pizza & Pasta. Pizza cooked in wood-burning ovens is featured. This is a favorite informal hangout for the locals.

AMENDMENT

The airport in Mannville, New Jersey, is now called Central Jersey Regional. It used to be called Kupper. The ID is 47N. Pizza & Pasta is still there and still a good place to eat. The airport runway is being repaved as I write and should be completed this week. The airport has new owners, and many other improvements have been done or are planned.

Marlboro (Marlboro - 2N8)

Sloppy Joe's 

Sloppy Joe's, 908-591-9805, is on the field. The food is typical airport fare.

Millville (Millville Muni. - MIV)

Flight Line Restaurant 

Saturday and Sunday breakfast is very popular at the Flight Line Restaurant, 609-825-3200. This historic exmilitary airport has a welcome small-town atmosphere. Taxi up to the fence right in front of the restaurant. Sausage biscuit with gravy is a specialty. The food is reasonably priced. The restaurant has approximately 15 to 20 tables.

Mount Holly (South Jersey Reg. - 7MY)

Airfare 

South Jersey Regional (7MY), which is roughly three miles from N14, which explains the right traffic on 1 at N14 and 26 at 7MY, has a lunch place right inside the terminal. This cafe caters to the burger and chips crowd. The price for a burger/chips/coke here would be around $7.00.

AMENDMENT

I'm only a lowly student pilot, so I've only physically flown into this place twice, but I'm fond enough of it to drive there when I get the opportunity. The new owners have done quite a bit to clean this place up, and the food is excellent and reasonably priced. Hamburger? Go for the filet!

Ocean City (Ocean City Muni. - 26N)

Airport Restaurant

Ocean City, New Jersey, has a small cafe on the field which serves good breakfasts. They are very busy on the summer weekends. The airport is close to the beach, and a short cab ride gets you to the boardwalk where there is plenty to do.

Pittstown (Sky Manor - N40)

Sky Manor Cafe

The Sky Manor Cafe is a small grill. The food is actually pretty good, but the hours of operations are limited. (908-996-3442)

AMENDMENT

They have a buffet on Friday nights that is pretty good. We go there two or three times a month. Good home-style cooking at an attractive airport!

Trenton (Trenton/Mercer - TTN)

The General's Quarters

The General's Quarters is on the second floor of the main terminal building at Trenton/Mercer Airport (TTN). The restaurant is a fine dining establishment that does not rock and sock you. Dinner for two runs about $35.00 - $40.00, if you skip dessert. There is an excellent view of the airport grounds. Trenton has two loooong runways, with an ILS tower. The Yardley VORTAC (ARD) is about three miles from the end of runway 6. As airport restaurants go, this is a definite five. TTN is a busy airport during the week.

West Millford (Greenwood Lake Airport - 4N1)

Connie's

Greenwood Lake Airport's (4N1) Connie's offers decent food for lunch every day. Breakfast is available on the weekends. The winds from the ridges provide a good refresher for crosswind approaches.

Alamogordo (Alamogordo -White Sands Regional - ALM)

Golden Corral 

Presently, there is not a restaurant on the field. Ed's Flying Service, located in the terminal building, usually has a courtesy car available for the usual fast food places, Golden Corral, Denny's, etc. There are also several steakhouses and Mexican restaurants.

Belen (Alexander Muni. - E80)

The Airport Deli 

Belen, NM (E80), has a great deli upstairs. Carolyn Parker makes some of the best breakfast burritos around, cinnamon rolls are homemade and unsurpassed, and the sandwiches are the best! They are open only for breakfast and lunch, Tuesday - Sunday.

Carrizozo (Carrizozo Muni. - Q37)

Outpost Bar & Grill 

Carrizozo, NM (Q37), has a great restaurant in town — about a mile walk from the airport — that makes great green chile cheeseburgers! The FBO will direct you there.

Farmington (Four Corners Airport - FMN)

Senor Pepper's Restaurant 

We flew round-trip from NC to CA this summer and stopped twice at Four Corners Airport, in part because of the really great Mexican restaurant that is on the field there. This is some of the best country in the USA for sightseeing by plane.

AMENDMENTS

KUDOs for Four Corners FBO. We were a flight of four Army UH-1s on a cross-country flight. Great restaurant! If you stop in, tell them thanks from the Joes at Ft. Riley.

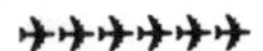

Senor Pepper's in Farmington is a great spot for true Mexican food! We fly to Senor Peppers from Santa Fe as often as possible. Great folks at the FBO; they'll roll out the red carpet for you, literally! If you're within a couple of hours flight of Farmington, make the trip. It's well worth it.

Hobbs (Lea County - HOB)

Wallace's Frosty Top

Flying into Hobbs? Stop and make use of the courtesy car to go to Wallace's Frosty Top for an excellent BBQ plate at a price less than $9.00. The FBO will provide directions.

Las Cruces (Las Cruces Intl. - LRU)

Crosswinds

Good home cooking is available at Las Cruces International. The Crosswinds is definitely worth a stop. It's located in the airport management building. I had a turkey sandwich last week that was a killer. The hours are 7:00 AM to 2:00 PM daily.

Lordsburg (Lordsburg Muni. LSB)

Mexican Restaurant

We stopped in Lordsburg for fuel. The lady at the FBO recommended a small Mexican food place in town. It was a very small, family-owned place with great food. I don't remember the name of the restaurant, but the FBO could tell you. It is only about three miles from the airport by courtesy car.

Buffalo (Greater Buffalo Intl. - BUF)

Italian Village Restaurant 

We were stuck in Buffalo, NY, waiting out weather on our way back to Cleveland. We asked about any good restaurants around, and a local suggested a new restaurant on Wehrle Drive called the Italian Village. We walked about 15 minutes to the west of Prior Aviation and found it. Remember those family restaurants in the late sixties? Pink walls, velvet wallpaper and old mints at the cash register? This must be the only one left. The FOOD was GREAT! The waitress was a straight-old-country and we had a GREAT time laughing with her. She even let us jam on their piano/drums combo by the bar! We're even gonna' make a special trip in just to visit again!

Farmingdale (Farmingdale Republic - FRG)

The 94th Aero Fighter Squadron 

My favorite fly-in restaurant is at FRG, Farmingdale-Republic Airport, on Long Island, NY. It's a WWII-decor restaurant with an excellent selection of food suitable for steak lovers and vegetarians alike. Wood-burning stoves are included, for warming up after those cold winter flights. The airport is between the New York Class B and C airspaces and has a $2.50 landing fee. The meal will typically cost about $40.00 for two, so this makes for a great place to take your first fly-in date. Ask to park on the main ramp but out of the way to the side. Then you just walk through the terminal and about one hundred feet and you're there. Highly recommended!

AMENDMENT

I'll second the motion that this is a great place. The 94th Aero Fighter Squadron restaurant has WWII aeronautical decor. The food is excellent. The menu selection is enormous. This is not your usual fly-in. This is a good place to impress a date or business associate with the benefits of a pilot's license.

Glens Falls (Glens Falls Muni. - GFL)

The Coffee Shop

My favorite airport restaurant is Glens Falls, NY (GFL). It is named simply The Coffee Shop, sometimes known as Tessies. The food is typical diner fare. The restaurant is in the old terminal building. The hours are limited so call first, 518-793-6359. GFL is located in the foothills of the Adirondacks near beautiful Lake George, the queen of North American lakes, and just north of Saratoga Springs, NY. There is an ILS approach if you need it and 5,000 feet of smooth, paved runway.

Hudson (Columbia County- 1B1)

Meadowgreens

Meadowgreens has a great Sunday brunch buffet from about noon to 2:00 PM. Park your plane on the east taxiway, push it off to the side so others can pass, and walk across the grass to the building abutting the airport. Rhinebeck is only about 10 miles to the south.

Ithaca (Ithaca/Tompkins County - ITH)

The Landing

I recommend The Landing, formerly Brian's Landing, at the Ithaca/Tompkins County New York Airport (ITH). You can easily park at the Taughannock Aviation FBO and walk to the old terminal building.

Mill Brook (Sky Acres - 44N)

Windsock Cafe

This is one of the best GA friendly airports near New York City. It is located in the beautiful Hudson Valley. Henry Kading keeps the weekend coffee shop open from 8:00 AM to 4:00 PM and serves up good food. If you get him talking airplanes, you might have to wait a bit for your food. The atmosphere and mountain view make this stop worth it.

AMENDMENTS

Henry Kading has retired. He left the restaurant in capable, and a lot prettier hands, I might add. The Windsock Cafe is now serving ultragood pies and cakes for finishing off that light lunch.

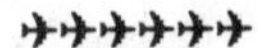

The Windsock Cafe at Sky Acres Airport is bustling with activity on the weekends now that the new runway is completed. Lee and Ann, the two very pretty operators of this restaurant, offer up a nice blend of home cooking and friendly atmosphere. With tables a mere 10 feet from the parking ramp, it is easy to see why this restaurant is one of the area's best for pilots and pilot wanna-bes.

Montauk Point (Montauk - MTP)

Crabby Cowboy

The Crabby Cowboy has opened recently at the airport, across the street, about a one-minute walk. Nothing fancy, but they have indoor and outdoor seating on the docks, on the water. The beach is a short walk away, just over the dunes at the end of the runway. The operator at the FBO will give you a ride to the beach in a courtesy van. BE WARNED: there is a $15.00 landing fee. The restaurant is open for breakfast, lunch and dinner seven days a week. The menu is VERY limited, offering only five items for breakfast, and burgers and such for lunch and dinner. Walk to the counter to order, and pick up paper plates, plastic forks, etc. All in all, this is a good place to get something to eat and spend an afternoon at the beach.

Montgomery (Orange County - MGJ)

Airport Cafe

Good food and reasonable prices. An aircraft dealer on field makes it interesting to browse and dream of bigger, faster and $200 burgers.

AMENDMENT

My wife and I just had lunch there. Great food — the onion rings are the best. Shirley, the owner, and operator is very nice and accommodating. You can taxi right outside the back door. Well worth the trip. Three runways for cross winds too!

Poughkeepsie (Dutchess County - POU)

Woronock House

A short 45 miles north of New York City you'll find the home of my all-time favorite $100 hamburger, the Woronock House. Don't ask where the name comes from; no one knows. I can tell you this though, the atmosphere is charming, the menu is extensive, the prices are cheap by New York standards, and the food is delicious! The restaurant also serves dinner, and on weekends they have live music. Since Dutchess County Airport is tower controlled, dinner flights are definitely doable. Since it's so well maintained, you can fly in year-round. When I flew up a few days after the Blizzard of '96, the runways were clean as a whistle.

Flying north from New York City the sightseeing is, in a word, fantastic.

Take the Hudson River route and you'll fly over the Tappan Zee Bridge, West Point, Bear Mountain State Park, and Sing-Sing Prison. If your departure airport is further south, then you can take the Hudson River exclusion north and circle the Statue of Liberty and check out the New York skyline — from an altitude that's about half as high as some of the buildings. New York never looked so good.

Once at Dutchess, park near the departure end of runway 24 near the Exxon pumps and the pilots lounge. There's a gate to go through and a short walk of about 100 yards and you're in "hamburger heaven." This place is highly recommended. Try the Cobb salad, it's great.

AMENDMENT

As one who has lived in the Poughkeepsie area for the past 20 years and had many lunch and dinner servings at the Woronock House, I can recommend it, but I don't think it is worthy of five hamburgers. The reported restaurants and others close by to Sky Acres (44N), Columbia County (1B1) and Montauk (KMTP) are better than the Woronock House.

Rhinebeck (Rhinebeck Aerodrome - PRIVATE)

Airshow Concession Stands 

Rhinebeck runs airshows every Friday, Saturday and Sunday from about Memorial Day through Labor Day. The airshow skit is a kind of Dudley DoRight action adventure with the Red Baron in there, too. Lots of old biplanes are flying and on display. The airshow starts around noon and goes until about 4:00 PM. Great fun to watch!

They tend to discourage using Rhinebeck as your landing spot since it is not a great field. It is only about 1,900 feet, with hills, grass and a few pieces of rock ledge showing through. You will find the runways "Xed" out now. Bottom line is, call for permission.

The alternatives are Red Hook, where they charge $10.00 for parking and then $3.00/person each way for a taxi. You can also land in Kingston, which has free parking but the taxi costs more!

Rhinebeck does have concession stands. Picnic benches and a grassy area make for a nice picnic if you buy or bring your own.

It's a great time if you can make it!

Saranac Lake (Adirondack Regional Airport - SLK)

Airport Cafe 

The restaurant at the Adirondack Regional Airport is open seven days a week from 7:00 AM until 3:00 PM. The menu has almost everything you could want.

Wallkill (Kobelt Airport - 10N)

The CAVU Restaurant 

The CAVU Restaurant is open for lunch and dinner every day but Monday. The food is very good and the prices are reasonable.

Westhampton Beach (Gabreski - FOK)

Belle's 

This is a friendly cafe located on an airport with a huge 9,000-foot runway. They offer a good variety of specials that change daily. I have also had the occasion of stopping in on a Sunday afternoon when they had live music. I'm not sure if this is a regular feature.

AMENDMENTS

The restaurant at Gabreski Airport, Westhampton Beach, NY, is Belle's. Yes, they do have music there every Sunday. They also have West Indian and Cajun specialties, which are, to say the least, unusual for an airport restaurant. It is open for breakfast and lunch.

✈✈✈✈✈✈✈

We just came back from Westhampton, NY. Everyone was right. The restaurant is superb. The only problem we had was deciding what to order. I wanted the ribs, but 10:00 AM was just too early. There wasn't anything coming out of the kitchen that I would turn down.

White Plains (Westchester Cnty. - HPN)

Wings of Westchester 

White Plains and gray ones and red ones...all in two queues for the active runway. This has the style of a typical New York coffee shop diner, where you'd half expect the waitress to call you "Hun." While enjoying the standard coffee shop fare, look around you to see what people are having for dessert. The rice pudding is really, really good. There is a chance that this may give way to the —REAL— $50 hamburger, available under neon signs at other major airports. HPN is under reconstruction in order to handle the new Lockheed Constellations and heavy jets rumored to be in use by the airlines. Let's hope that Westchester's officials still appreciate the restaurants of an earlier era.

HPN has two active runways. The smaller one generally has no traffic.

AMENDMENT

There's more to HPN than the coffee shop. The restaurant, Wings Of Westchester, was good in the old Quonset huts and remains good in the new jetport setting. HPN is schizophrenic: If you fly in your L3 with a handheld radio and ask to park at general aviation, they'll direct you to a ramp 100 yards from the 737s, MD80s and ATRs. The restaurant is on the second floor of the new terminal between the coffee shop and the bar. No jetways block the view. I'd give the whole show four burgers.

Arriving at HPN is de facto Class C. You need to coordinate with the approach facility, NY Approach, over the city of Stamford, CT, or over the Hudson River. Typically they want you to fly to the field from the Tappan Zee Bridge. Departure is standard Class D procedure.

Skytop Restaurant 

Cheeseburger In Paradise!

You owe it to yourselves to reevaluate the facilities at the new Westchester County Airport. The new Skytop Restaurant has a new bar, new diner-eatery, new, new, new. They haven't yet taken the backing off of all of the new Plexiglas! Despite the war between my nostalgia for this WWII Quonset-hutted facility and the new jet-age, I have to vote…I like 'em both! Wave me off and let me land, all at once!

Charlotte (Douglas International - CLT)

G's Burger Joint

G's Burger Joint is at CLT, close to Signature. They have good food at low prices.

Fayetteville (Grannis Field - FAY)

Airport Cafe

I fly back and forth from south Florida to Leesburg, VA, a lot. Fayetteville is an easy second fuel stop in a 172 or similar airplane. This is definitely no more than a two burger rating, standard airport terminal fare.

Gibson (Stanton's BBQ - Private)

Stanton's BBQ

Stanton's BBQ is a real fly-in restaurant. About 20 steps from the plane parking is the large red restaurant building. It's a private field -- not on the sectional. The coordinates are N 34 . 43 . 9 - W 79 . 40 . 0. If you're flying from the Florence, SC, to Sandhills, NC, VOR route, you'll fly very close. It's near the town of Gibson on the NC/SC border, just southwest of Fayetteville. The restaurant is immediately adjacent to the field. Be sure to tell the hostess you're a pilot. They have a special room for pilots and families only. They're open every day for lunch, including weekends. The fare is basic country cooking; the BBQ is great. This is a great place to take the family. The strip is smooth grass, about 2,200 feet long with easy, clear approaches but it is unlighted. The Unicom is 122.9. Call for "BBQ traffic" and you'll get a friendly response.

AMENDMENT

Stanton's BBQ is actually in SOUTH CAROLINA, but three miles SW of Gibson, NORTH CAROLINA. Since they don't put that nice state line on the sectional, most of us truly don't know which state we're in anyway! The telephone number is 803-265-4855.

Hickory (HKY)

The Runway Cafe 

The Runway Cafe located at Hickory, NC, has reopened under new management. They are open Monday through Friday 8:00 AM to 5:00 PM. They make homemade soup every day. It is run by a husband and wife team, who take a lot of pride in their restaurant. Worth a try if you are in the area. The restaurant is located in the terminal building.

Jefferson (Ashe County - NC67)

Shatley Springs 

Shatley Springs, eight miles from Ashe County Airport, is the ultimate in Southern home-style cooking. Just call the restaurant and one of the staff will come to the airport and pick you up. This place is the ultimate Southern all-you-can-eat experience.

Kenansville (Duplin County - DPL)

The Country Squire 

Land at Duplin Co. (DPL), for a really unusual restaurant. The FBO there is really helpful and maintains a courtesy car. Go about three miles to The Country Squire. It has a large and varied menu as well as a five-star atmosphere. The restaurant resembles a very old log cabin. They have all alcohol permits and a few quaint overnight accommodations if you wish to join your passengers. It is truly a unique place.

Liberty (Causey Airport - 2A5)

Fran's Front Porch 

Located at Causey Airport (2A5) in Liberty, North Carolina, it is one of the nicest airport restaurants I've encountered. Located in a beautifully restored country mansion adjacent to the airport, Fran's Front Porch serves nothing but the finest Southern home-cooked meals. The Sunday morning brunch is the most popular. Served buffet style with a choice of three or four main dishes and all the vegetables, fresh-baked bread, salad, and desserts you could ask for. The runway is narrow, but hard-surfaced, and recently lengthened from 3000 feet to around 5000 feet. Well worth checking out!

AMENDMENT

Fran's Front Porch specializes in home-cooked meals and terrific desserts. It's right across the street from a small but very busy airport. There is some traffic between the airport and Fran's, so word must be out. When I say "traffic," I mean foot traffic. The distance is probably no more than 500 yards between the terminal and Fran's front door. I don't know if you can really call it a mansion. It's just a big farm house.

Mt. Airy (MWK)

52 Seafood Restaurant

There is a nice restaurant on the hill behind the FBO. We arrived at 3:45 looking for something to eat only to find the place closed. As we turned to leave, Jimmy, the cook, came out and told us the hours were from 4:30 PM - 9:30 PM Thursday through Sunday, but if we didn't mind him waiting on us, we were welcome to come in.

The rest is history! The food and service were great. When you stop by for dinner, tell "Jimmy the cook" you read about him.

Ocean Isle Beach (60J)

The Dawg House

The Dawg House is a very nice, short-order-only cafe 100 feet from the tie-down area. Highly recommended!

Ocracoke (Ocracoke Island - W95)

Howard's Pub

On the island of Ocracoke on the North Carolina Outer Banks, there's a wonderful pub called Howard's Pub. The food is good but not outstanding. Your passengers will love the 250+ different beers they pour. Stay overnight so you can indulge. Rate it about 3.5 burgers.

The Back Porch

For better dining in the evening try The Back Porch, lots of seafood prepared with herbs grown in the back yard. You will have to stay overnight, since the airport (W95) is not lighted. I'd give The Back Porch 4.5 or 5 burgers.

The Atlantic Cafe

The Atlantic Cafe is our favorite restaurant on the island. Located about a mile from the airport, it is one of the first restaurants encountered when walking from the airport toward town.

Being Midwesterners starved for seafood, we were delighted by all of the fresh products of the local waters. The kitchen staff pushes beyond customary offerings with twists of originality. The wines are primarily Californian but a couple of terrific French labels save the list from mediocrity. Dinner for two runs around $45.00.

If your flight is IFR, polish up the NDB navigation. The airway and approach are based on the longwave.

Pie in the Sky

If you get to Ocracoke, Pie in the Sky, almost across the street from Cafe Atlantic, has great draft beers. Guinness, Bass, Harp and many others are available. Ask for a black & tan, Bass Ale with a layer of Guinness Stout gingerly poured on top. They also serve salads, subs and some of the best pizza you ever put in your mouth. The homemade dessert pies are also incredible.

Roxboro (Person County - TDF)

The Homestead Steakhouse 

Located only a couple of miles from Person County Airport in North Carolina, Homestead offers the most sumptuous buffet beginning each Sunday at 11:00 AM. The line is packed to overflowing with good country fare. The sideboard is loaded with cakes, pies, and puddings, the likes of which are not found within a thousand miles. Just have the FBO call and they pick you up in a van. For $6.50 a head, it's the best deal in the universe. Pack lightly, you may over-gross the airplane on the return trip.

AMENDMENT

I just got back from Roxboro, Person County Airport. The Sunday buffet at The Homestead Steakhouse has been discontinued. They open at 4:30 PM on Sundays and have a great selection of steak and chicken dinners priced around $10.00 to $15.00, with a few entrees at $5.95, including potato and coleslaw. A blooming onion goes for $3.95. Very friendly FBO that lent us the courtesy car to go into town of Roxboro. Nothing is open on Sundays in town. Very nice terminal with snack machines and a weathermation.

Rutherfordton (Rutherford County - 57A)

57 Alpha Cafe 

Great breakfasts and lunches! In the Heart of the North Carolina mountains. Five tables, so get there early on Saturday. It is not open on Sunday. Occasionally during the summer, dinner is served, with Ron, the owner and chief cook, playing his guitar and singing. Regular hours are about 10:00 AM - 3:00 PM Tuesday through Saturday. Luncheon specials are chicken, taco salad, hot dogs, and great chili. You are always made to feel welcome at 57A!

AMENDMENT

The 57 Alpha Cafe deserves another burger on the rating scale. It's located at Rutherford County (57A), in North Carolina. Ron will always fix you a great lunch. On a Saturday the place will be packed with pilots from all over the region who make it their only flight of the month. He's open until 3:00 PM daily, but closed on Sunday and Monday. Ron has a varied menu and he prepares it fresh. The homemade banana pudding is great, as is the fellowship. It isn't large, just a real nice place for pilots and their passengers.

Siler City (Siler City Airport - 5W8)

Great Bar-B-Q 

Siler City Airport (5W8) is 30 nautical miles west of Raleigh and south of Greensboro. There is no on-the-field restaurant, but you can always borrow a courtesy car for a quick trip to town for great BBQ or family-style seafood. You can also rent a car on the field on a weekend for $20 and drive 18 miles to the Asheboro Zoo! The field is lighted and 5,000 feet long, with nice, easy approaches.

Grand Forks (Grand Forks Intl. - GFK)

The Crosswinds Cafe 

The Grand Forks International Airport has a closely guarded secret, the Crosswinds Cafe in the airport terminal building. The food is great, all homemade, and the staff is friendly and helpful. Feel free to stop by, and don't forget to ask about the weekly special. Of course, no trip to Grand Forks International Airport is complete without stopping by the airport fire and rescue station to say "Hello" to Little Ricky, the airport's resident cat. He dines on the $100 mice!

Minot (Minot Intl. - MOT)

Airport Restaurant

The Airport Restaurant is a pretty standard large terminal kinda' place. What makes it worth a stop is the airport. It has nice long runways, so density altitude is never a problem, even if you're hauling a major moose back from Canada. The field elevation is only 1,700 feet anyway! Any quick repair your ship needs can be done at this full-service airport. The restaurant will provide a large coffee container if you ask Carol. She's the cheery brunette!

Akron (Fulton Intl. - AKR)

Cafe Piscitelli 

Cafe Piscitelli is located in the old terminal bldg. There is a ramp in front of it, the old customs ramp, for A/C parking. It has Italian food with a nice atmosphere, cloth tablecloths and napkins. Plan around $12.00 per main dish. The food is excellent.

AMENDMENT

Returning to Wisconsin from NY, my wife and I stopped to refuel and have lunch at the Cafe Piscitelli, located in the old terminal building at AKR. We were pleasantly surprised by the quality, price, and atmosphere of this excellent Italian restaurant adjacent to the ramp!

As we walked through the gate from the ramp feeling rather beat, and frankly not in the best of spirits, the owner greeted us with a wide smile and a hearty, "Welcome to my restaurant, please come in! Enjoy your lunch." That is exactly what we did. The pastas were excellent — deliciously spiced with some of the best meatballs I've ever encountered. Service was top-notch, and everything was very reasonably priced. Two pasta dishes and one Italian beer, for my wife of course, cost only $17.00. Hey, the beer was four bucks.! We left feeling renewed, and vowing to return anytime our route takes us nearby. Frankly, if there's a five-hamburger rating available, I'd give it here. Exemplary in value and style and hopefully a new breed for airport digs.

Batavia (Clermont Cnty. - I69)

Skyline Chili Parlor 

OK, this isn't really in Clermont Co. Airport, but it is a fun way to spend a lunchtime. Fly into Clermont Co., home of Sporty's Pilot Shop, in case you didn't know; wander through the shop and do a little spending. Save some $$, however, because you can use their courtesy car to drive about seven miles to a Skyline Chili Parlor. Order the "5-way" and don't ask. If you haven't had Cincinnati chili before, it is sweet and spicy and to die for.

Sporty's Pilot Shop 

Every Saturday at Clermont County Airport (Batavia I69) Sporty's serves up free hotdogs, melts and brats. Dogs are served from about 11:30 AM until 2:00 PM.

Bluffton (Bluffton Field - 5G7)

Denny's

Denny's is on the field, hotels too. Not really a pilot hangout, but hey, it's a Denny's.

Carrollton (Tolson Airport - TSO)

Granny's

This restaurant is one of the best I have ever eaten at. If you want family-style, down-home cooking, this is it! The homemade pies are excellent. The fudge brownie with ice cream is fabulous. They have great taco salads, hamburgers, etc. Place your order over the Unicom!

Blue Bird Restaurant 

Just down the hill and through the woods a couple hundred yards – well-marked and beautiful — from Granny's, is as delightful a place as you will find close to an airport in Ohio. Built in the mid-1800s, the Blue Bird Restaurant offers an unusual menu and a setting that will have you returning with someone special! The prices are good, food excellent and the environment great! A well-stocked gift shop is located next door. They specialize in bluebird houses unlike any you will see elsewhere.

AMENDMENT

We visit TSO almost weekly, whenever the weather is decent for weekend-warrior-type pilots. The food is always excellent. The desserts are considerable portions, as is everything else. There is usually a special of the day that will not leave room for dessert unless you are a big eater. The fuel prices are usually among the best in the tri-state area, thanks in part to the local Pilots' Association. Carroll County (TSO) also has one of the best Pilot Shops, Lamp Aviation, run by Leon Lamp, and they also are very handy with avionics and mechanical needs. Announce yourself greater than five to ten miles out, and listen to the Unicom 122.7, as you will find plenty of traffic on the weekends.

Cincinnati (Lunken Field - LUK)

Wings 

My favorite place to get a $100 burger is The Sky Galley at Lunken Field in Cincinnati, OH.

AMENDMENTS

The Sky Galley has changed hands and is now known as Wings.

Appetizers, The Preflight, run $6.00-$7.00

Soups, The Run Up, go for $2.00-$2.50
Burgers, In the Pattern, $5.95-$6.45
Meals, In Flight, up to $12.95 (steaks, shrimp, etc.)

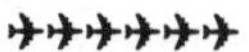

LUK's Wings deserves at least a four-burger rating. You can taxi up to the Wings door. This summer, Wings is featuring 1.5+ pound lobster dinner for under $18.00 on Friday and Saturday nights (corn-on-the cob, redskin potatoes ...).

Cleveland (Burke Lakefront - BKl)

Hornblower's in Cleveland, OH, is within walking distance of Burke Lakefront (BKL). It is located in a boat basin between the terminal and the adjacent Rock & Roll Hall of Fame. Hornblower's features a daily menu, including soups, sandwiches, seafood, salads, and other full dinners.

AMENDMENT

I just came back from Cleveland, landed at Burke Lakefront and checked out Hornblower's. GREAT atmosphere, GREAT waitstaff, and really, REALLY great food! They had a good jazz combo playing on Friday night!

Columbus (Bolton Field - 214)

JP's Ribs

A great place to eat with a fun atmosphere is JP's Ribs at Bolton Field (2I4), just southwest of Columbus. On weekends biplanes usually perform aerobatics.

Coshocton (Richard Downing - I40)

Roscoe Village

A good fly-in spot in Ohio is Richard Downing Airport near Coshocton, OH. The airport is located on top of a ridge above Roscoe Village. This is a historical canal town with small shops selling antiques, quilts, ice cream, etc. You can even ride a canal boat. There is a motel and several restaurants. Call the shuttle van from a booth near the airport terminal building for the short ride down the hill. The airport sells 80 and 100 octane avgas at a reasonable price. The place is worth four hamburgers in my book!

Dellroy (Atwood Lodge - OI56)

Atwood Lodge

Atwood Lodge has a 3,100-foot grass strip that is usually well maintained. There is a phone near the aircraft parking area that calls a van from the Lodge at no charge. Food is good and if you ask for a window table, you'll have a great view of the lake.

Mansfield (MFD)

The Flying Turtle

The Flying Turtle is in the old terminal building, which has been completely remodeled in a very pretty oak, green, and burgundy. The view is excellent; large windows face the crossing runways. The restaurant is open from 7:00 AM to 7:00 PM seven days a week. The menu includes soups, salads, burgers, Reubens, chicken, desserts, and appetizers. They have a good children's menu.

The decor is definitely aircraft. My favorite is the tabletops. They took sectionals from all over the world and set them into the tabletops. Models, parts, and other memorabilia hang from the ceiling and walls. A scanner pipes in ATC frequencies. TV monitors show aircraft videos.

There is a pilot shop in the restaurant which has jackets, shirts, books, and videos. They told me they plan to sell sectionals and NOS approach plates in the future. There is a WSI weather terminal in the pilot shop and some leather chairs.

Middletown (Hook Field Muni. - MWO)

Frisch's Big Boy

Frisch's Big Boy is adjacent to the field, and there is a special ramp for restaurant parking. Big Boy restaurants have a noteworthy breakfast bar.

Mount Victory (Elliot's Landing - 84)

Plaza Restaurant

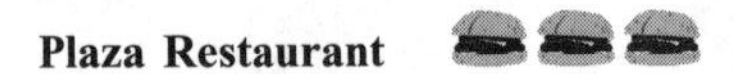

Located adjacent to this 2,750-foot turf strip is the Plaza Restaurant, which has great food at very reasonable prices. For home-cooked meals, large portions, and excellent menu selection, this is a tough one to beat.

Ottawa (Putnam County - OWX)

Red Pig Inn

The Red Pig Inn has fantastic barbecue! This is a five-time NW Ohio Rib-Off winner! The restaurant is off the field but offers free courtesy transportation. Call them at 419-523-6458 to be certain it is available. The menu is varied, with steaks, seafood, barbecue, pasta, salads and much more. A casual dining atmosphere is punctuated with a great staff and friendly service. Their fabulous Sunday brunch is a must for weekend flyers!

New Philadelphia (Clever Field - PHD)

The Hangar Restaurant

This restaurant has really good steaks. I haven't been there in a year or so, but the food is always good. This place is more of a dress-up place than your normal burger hangout, but not so dressy that you would feel uncomfortable in a nice pair of jeans and a dress shirt. It's a great place to take a date. It also has breakfast, although I haven't eaten there for breakfast. A little more pricey than Granny's over in Carrolton.

Port Clinton (Port Clinton - PCW)

Nate's Restaurant 

Nate's Restaurant has a variety of delicious entrees served in a casual atmosphere. Call Nate from the airport and he'll pick you up. His restaurant is about one mile away.

Phil's Inn 

Phil's Inn is about three miles from the airport and specializes in spaghetti and other Italian foods. Call them from the airport and they'll pick you up.

Portsmouth (Greater Portsmouth Reg. - PMH)

Skyline Family Restaurant 

A nice diner located right on the ramp, they have been known to have warbird fly-in's.

Put In Bay (Put in Bay - OH30)

Skybar 

On the north end of Put-in-Bay Airport (OH30) is the place the local residents go for a good perch dinner. The flight is interesting, as the airport has Perry's Monument to the north and the runway has a dogleg in the middle. Park in the grass at the terminal and walk to the north end of the runway to the restaurant. Departures after sunset are now prohibited!

AMENDMENT

Put-in-Bay has a nice new runway and ramp; no more dogleg in the runway and the ramp is all asphalt. You can eat at the Skybar at the north end of the field or take a walk one mile north uptown to the docks. The nightlife is the wildest, so plan to stay over as you can't leave after dark legally.

Sandusky (Griffing-Sandusky - SKY)

Cherokee Inn 

There is a small restaurant on the field, and it is a nice place to go on a Saturday or Sunday morning.

AMENDMENT

The Cherokee Inn, Sandusky, OH (SKY), only serves breakfast and lunch. They close at 4:30 PM.

Sebring (3G6)

The Landing

The Landing is located in Sebring, OH, which is close to Canton. The restaurant is very busy, as it has been ever since I've eaten there. There are parachutes and flight jackets on the wall, definitely an aviation atmosphere. The food is delicious. Thankfully, the variety changes often. It isn't a very busy airport, so you can eat without too much aircraft noise. There is a nice lounge building adjacent to the restaurant.

Urbana (Grimes Field - I74)

Airport Cafe

Nice diner-style restaurant on the field in the FBO building. It doesn't get much closer than this one. It's good for breakfast and lunch. They even had an Ercoupe flyin last fall!

AMENDMENTS

We've eaten here several times and it's great food, served in a clean, bright diner atmosphere at very reasonable prices. It has the typical aviation-oriented meal names, but they certainly don't become carried away with it. It's so good the locals come out for Sunday breakfast!

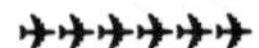

It's just as good as the previous reviews say; not fancy food, but good, and well worth the trip. There were plenty of people from Urbana who drove there, and while we were eating, seven more planes landed for breakfast. Word gets around about a good thing. Get there early. They are open from 8:00 AM to 3:00 PM but stop serving breakfast at 11:00 AM.

Wauseon (Private)

Smith's Family Restaurant 

Smith's Family Restaurant is immediately adjacent to a private 2,700-foot turf strip in golf course condition. Land at your own risk. The coordinates are 41 35.37N 84 08.71W. It's location, at exit 3 on the Ohio Turnpike, makes it easy to find. The field elevation is 785 feet. Fly a right-hand pattern to 9 or left to 27. It is used by a lot of pilots. The restaurant is open 7:00 AM to 10:00 PM every day except holidays. Call 419-335-4896 just to be sure. The food is great.

Ada (Ada Muni. – AHD)

Bob's BBQ

Bob's BBQ is located about a half-mile from the terminal building in Ada, OK. Walk down the airport's driveway and turn left when you reach the highway. Bob's is about a quarter-mile ahead on the left. Be careful — there's no sidewalk or guardrail between you and a fast divided highway.

Bob's is classic BBQ from the word go. It is a really friendly place with fantastic food. They have a Big Bob sandwich that is gargantuan! The BBQ comes in all different varieties, shredded and chunked beef or pork, plus grilled chicken breast, on sandwiches and combination plates. They've always got fresh homemade pie for dessert and the drinks are bottomless. Very reasonably priced, a sandwich, drink and pie costs about $7.00. Enjoy!

Afton (Grand Lake Regional Airport - 3O9)

Crow's Nest

Shangri-La reopened last summer after being closed for a period of time. The new owner has gone all out to make life comfortable for those who visit. There are several places to eat on the island, and he has a limo to take you any place you wish to go. We go to the Crow's Nest.

AMENDMENTS

Shangri-La Airport (3O9) has had a name change. It is now Grand Lake Regional Airport. Nothing else has changed, except they are adding a fast-food counter in the terminal. They still provide complimentary limousine rides to all Monkey Island locations.

✈✈✈✈✈✈✈

On the north end of the island, at Port Duncan, a restaurant called Crow's Nest more than fills the vacancy. A beautiful lake view and affordable prices are added features. The airport management still transports you all over the island as a courtesy. Great place to visit!

Ardmore (Ardmore Muni. - ADM)

The Runway Cafe

Need a great cross-country destination into a controlled field? The newly opened tower at Ardmore (ADM) has a terrific cafe right in the bottom of the building. The restaurant monitors the Unicom (122.95) and can give you one-stop shopping for fuel and fuel.

They feature freshly smoked turkey sandwiches among their wide selection. As you wait for your breakfast or lunch, you are encouraged to write your name and tail number on the wall in magic marker. A TV monitor rolls continuous tape of military flying video, from WWI to present. Don't forget to check out the old F-4 ejection seat in the corner. Outside, if you have the time, you can walk among the 747s United Airlines abandoned in their boneyard.

The approach is beautiful, with the airport nestled among the verdant hills and the horseshoe bend in a river. When we flew in on a Saturday, we were about the only plane at the field, but Ron said it's a popular training destination for planes out of Sheppard AFB during the week. Highly recommended! The Runway Cafe's phone is 405-389-5555.

AMENDMENTS

I was just at Ardmore (ADM) this weekend and the restaurant has changed ownership. It is now owned by a great lady who just happens to be married to the guy who runs the FBO. The food is standard airport cafe fare, but quite good with friendly service.

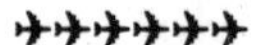

The cafe at Ardmore, OK, was truly wonderful. I went and was amazed by the friendliness of the people.

Bartlesville (BVO)

Murphy's Steak House

Murphy's is two miles from the airport. Don't worry. Two courtesy cars are available. This modest-looking place on the low-rent side of Bartlesville attracts hordes of hungry "Bartians" for lunch and dinner. You can sample a true Okie specialty: the "hot hamburger." Basically, it's grease-laden white bread topped with a lard-oozing beef patty, a monster heap of french fries, all smothered in rich brown gravy, roughly a gallon, with onions on top; a veritable heart surgeon's dream-come-true. Get directions from the fine crew at the FBO.

Cookson (Tenkiller Airpark - 44M)

SmokeHouse Restaurant

A must to try! Located in Cookson, Oklahoma, on the east side of Lake Tenkiller is the SmokeHouse Restaurant. This log cabin restaurant has excellent BBQ, but also specializes in homemade biscuits and gravy, steaks, fish and homemade apple dumplings. It is open seven days a week for breakfast, lunch and dinner. It's a great place to eat, but you better bring BIG appetites with you. You can experience the cozy atmosphere of the SmokeHouse by flying to Tenkiller Airpark, one mile SW of Cookson, which has a 2,600 foot x 75 foot turf runway with VASI lights. Runway 5 is the preferred runway for landing, with 23 the preferred take off runway. The telephone numbers are: SmokeHouse Restaurant 918-457-4134 and Tenkiller Airpark 918-457-3257.

Cushing (Cushing Muni. - CUH)

Silver Dollar Cafe 

Two-pound cinnamon rolls and steaks that hide the plate! Connections, however, were nonexistent. I ended up walking the 2-3 miles for a great reward. I had landed after dark and couldn't get any ride to town. There are now 12 Silver Dollar Cafe locations across NE Oklahoma, a minifranchise. A few are near airports. I plan to repeat this trip until I find another pilot-friendly location. Well worth the cross country, as if I needed flying to be any more fun. We're talking great menu.

AMENDMENT

Cushing Municipal is a pilot-friendly place, but there are no services after dark. There is a courtesy car available on the field that can be used to visit several downtown restaurants.

Duncan (Halliburton Field - DUC)

Phipp's Bar-B-Que 

Phipp's Bar-B-Que is a short walk out the gate and to the left, approximately 1/4 mile from the FBO. The smoked steak is the "best in Oklahoma." You can ask for the courtesy car for the short drive, but if it's not raining you won't need it. Well worth the trip!

Enid (Woodring Airport - WDG)

The Barnstormer 

Enid, OK, has a good restaurant on the field — homemade pies, good food, and a good view of the airfield. What could be better? Their telephone number is 405-234-9913. WDG is a nice airport with an NFCT. Fuel is available, as is a courtesy van. Make note that WDG is not far away from Vance AFB, which has lots of jet training traffic!

Goldsby (David J. Perry - OK14)

Hydes's Barbecue 

David J. Perry Airport is located just south of the South Canadian River and is about 13 miles south of Norman Westheimer Airport. The airport is a military surplus Navy training field that is currently owned by the town of Goldsby. The airport has two runways, 13/31 and 17/35. It is a very warm, friendly place. There is a very active volunteer group which oversees all activities at the airport.

Try Hydes's Barbecue for a treat to remember! Just a ¾-mile walk, you'll probably need it after the meal, to where you park the airplane.

Lake Texoma State Park (Lake Texoma - F31)

Lake Texoma Lodge 

I've always enjoyed the trip from Dallas up to Lake Texoma State Park (F31) to have their breakfast buffet. The food is good and the people there are very friendly and make you feel right at home. It's a short walk up from the airstrip to the Lodge, where the restaurant is located. Definitely worth the trip!

AMENDMENTS

More info. on Lake Texoma State Park Resort restaurant. Open from 0700 - 2000 Monday - Friday, 0700-2130 Saturday and Sunday, 365 days a year. Sandwiches in the $4.00 - $6.00 range, entrees $6.00 - $15.00, including catfish and country fried steak.

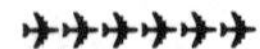

My Dad and I ventured out last weekend and were pleasantly surprised. We intended to try out Sanford's Steakhouse, about 700 yards west of the field and supposedly one of the best steaks around, but it was closed until 4:30 PM, so we went back to Lake Texoma State Park Resort. They have changed their menu and it was terrific! I had a very good chicken-fried steak. Dad had a terrific bowl of vegetable soup. We wasted our money on the coconut cream pie and didn't need the extra weight on takeoff!

Marietta (Love County - OK31)

President's Bakery 

You can't get a burger there. It's not even a restaurant. It's really more like a cookie factory!

Land on the grass at Love County (OK31) in Marietta, OK, and leave your plane next to the T-hangars. Go through the gate behind the hangars and cut through a narrow field to get to a farm road. Watch your step; there's a herd of cattle that frequents that field. Follow the farm road south until you get to a stop sign at a major road turn left; this road will take you under I-35. Keep walking until you see the sign, COOKIES. Total trip is about 1.5 miles. Do it on a nice day and you won't even notice. I believe the factory is called President's Bakery. There's a store at the factory that sells 4-5 lb. bags of broken cookies for $1.00 each. The store is open every day until at least 5:00 PM. Take enough to get the plane back to maximum takeoff weight and you'll never need cookies again!

Marietta (McGehee Catfish Restaurant - T40)

McGehee's Catfish Restaurant 

Located in Marietta, Oklahoma, McGehee's is nothing but a 2,250 foot x 30 foot uphill grass strip. Taxi to the end, hop out of your plane, and about 50 steps away is the finest catfish anywhere! They're open from 4:00 PM till about 9:00 PM. Check this place out! I've been there at least a dozen times. It's only about a 40-minute flight from the DFW area.

AMENDMENTS

McGehee's is a true "fly-in" restaurant. The approach is a little tricky, because if for some reason you slid off the active, you'd fall about 30 feet into the Red River. The takeoff is tricky from both directions. The south RWY has very rapidly rising terrain at the end of the strip. Worse is 35, which runs you straight into the dinner table. The waitress told me that there have been two incidents where an aircraft said hello to the old catfish building. One was a fatality. Could there possibly be another restaurant that has had an aircraft come to the table?

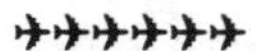

On such a beautiful day as today, it was impossible to stay inside, so I called a friend and we played hooky to check out a famous restaurant in Oklahoma, near Ardmore.

We flew from Arlington to McGehee's Catfish Restaurant in my 1956 172 against clear blue skies, with a 15 knot tailwind and smooth air at 4,500 in under an hour. The Loran field name is T40. It is just north of the Red River in Oklahoma, west of I-35 and barely east of the river. I-35 splits for a mile or so abeam the airport.

It is a lovely sod strip, cut into a small bank next to the Red River, and is lit for night operations. A good thing, as McGehee's is open from 1730-2130 daily, 1300-2130 Saturday and Sunday, and closed on Wednesdays. They accept Visa, MC, and Discover cards.

The runway slopes from south to north, so winds permitting a south landing and north takeoff are advisable. There is a road crossing about halfway down the strip, but it is a smooth juncture.

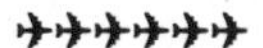

This is one of the best places for catfish in the South. You walk in the door and aren't overcome by a fishy odor. The restaurant overlooks their fish farm. Anyone who passes by an opportunity to eat at McGehee's is crazy!

McAlester (McAlester Regional - MLC)

Pete's 

Pete's serves some of the best Italian food in the state. This is attested to by all the autographed photos of famous persons adorning the walls. When our group (CAP) spends the weekend at McAlester, the unanimous choice for dinner is Pete's.

The airport is about eight miles from the restaurant, but Pete will gladly pick you up at the airport after you call. I recommend that you make reservations ahead of time. They will have your table ready when you arrive. I won't waste time on the menu. If you like Italian, you will be in heaven when you get there. Hint: Come hungry, Pete doesn't skimp on the portions. Pete's phone number is 918-423-2042.

Murray (Kyle-Oakley Field - CEY)

Rudy's

Rudy's in downtown Murray is world-famous for their hamburgers! Their Friday all-you-can-eat catfish is excellent!

Fly into Kyle-Oakley Field (CEY); someone will see to your ground transportation needs! The airport doesn't have a courtesy car any longer. It was never replaced after the wreck! There are a few transportation choices. Usually, either Covenant Aero, the maintenance operation on-field, or the airport manager will be glad to drop off and pick up. Otherwise, there is a city shuttle and a taxi service for reasonable rates. The trip to Rudy's is about 10 minutes.

Dutch Essenhaus 

The Dutch Essenhaus is home-cooking cafeteria-style. Bring a healthy appetite. You will not go away hungry! The trip to Essenhaus is about 5 minutes.

Oklahoma City (Downtown Airpark - 2DT)

Downtown Airpark Restaurant 

This is one of the oldest fly-in restaurants in our state. It has been in existence since 1952. It features good ol' home cookin' made by Kay, whose mother was the founder in '52. Fly in on Wednesdays for the Mexican food or Thursdays for chicken noodles. As usual, the cobbler is to die for. Add all that to the constant parade of Turbo Commanders that come for service and a jet or two and it makes for a wonderful dining experience.

Oklahoma City (Wiley Post - PWA)

Annie Okie's Runway Cafe 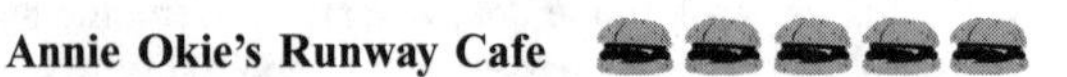

The cafe is located in the tower building at PWA. Good food and good company from 7:00 AM to 2:00 PM Monday through Saturday, and 9:00 AM to 3:00 PM on Sunday.

AMENDMENT

The Runway Cafe (Annie Okie's Runway Cafe) is a favorite weekend breakfast and lunch stop. They're open through the week, but the big crowds tend to convene during the weekends. Their omelets rule, and the burgers will most definitely make you want to come back. You'll see many familiar faces if you come more than once! The Runway Cafe is located in the terminal building at Wiley Post Airport (PWA), Oklahoma City. You can keep an eye on your airplane, and brag on it, from the restaurant atrium.

Overbrook (Lake Murray State Park - 1F1)

Lake Murray State Park Lodge - Fireside Dining 

What? Nobody's said anything about Lake Murray? You land on the park's own airstrip, just inland from the lake shore and adjacent to the 18-hole golf course. Walk up the steps to the clubhouse and call the Lodge. They'll send a van or car to pick you up, or you can walk the mile down the road through the Okie woods. It is a beautiful lodge, with a nice restaurant overlooking the lake. There are paddle boats and canoes nearby, or you might be able to persuade them to run you down to the big marina where you can rent a boat or jet ski or just go swimming. If you've got an amphib., know that the lake and marina welcome float and seaplanes.

AMENDMENT

For those unfortunate and discerning palettes that have not discovered the REAL place to dine at Lake Murray, let me describe. How about succulent mouth-watering prime rib or Cajun-style blackened fish? Want to be healthy? Any number of broiled fish, chicken, or shrimp dishes can fill the bill. A special occasion spent here is a must, whether it's a birthday, anniversary, or just a really romantic date.

Enough suspense, already. I'm talking about Fireside Dining, an actual sit-down-and-enjoy-it restaurant. One hint, bring time and your heavy wallet. The absence of green will offset the weight added after a truly awesome fly-in culinary experience. I usually allow an hour of dining time for this multicourse extravaganza. Budget about $20.00 to $25.00 per person without drinks, and if you're flying, please don't imbibe, for your safety and your loved ones.

Gary (the owner) and his staff always make every dining experience truly top notch. The numbers for all restaurants and other services are listed conveniently near the phone at the golf club. You may request the clubhouse attendant to call Fireside Dining for you. The number is 405-226-4070.

The restaurant is close enough to walk, but who can resist a ride in Gary's new Suburban, in which he will arrive within mere minutes of your call to personally greet you and your party.

Paul's Valley (Paul's Valley - F61)

Punkin's Bar-B-Que 

Punkin's Bar-B-Que is located in Paul's Valley in south central Oklahoma. The specialty is BBQ and catfish. I've had both and it's a tough choice. Punkin's is about three miles from the airport. Last time, they didn't have a courtesy car. They're working on it, but Punkin's will come and get you. The phone number for Punkin's is 405-328-2320. Their hours of operation are Wed. - Sat. 10:30 AM - 9:00 PM, Sun. 10:30 AM - 2:00 PM, Mon. and Tue. CLOSED.

Ponca City (Ponca City Muni. - PNC)

Enrique's Mexican Restaurant 

The restaurant is in the terminal building. The food is great and reasonably priced. My wife used to like to go there because it was the only place she knew where you could eat great Mexican food and admire your airplane all at the same time!

Haven't been there in a while, but the restaurant used to be closed on Wednesdays.

AMENDMENTS

I, too, am an avid fan of Enrique's at the Ponca City terminal. I have flown there many times on any excuse to enjoy the food. The chips are puffed when they are fried and are different from any others I have seen. Furthermore, it's nice to be able to admire your plane parked a few feet away while you dine. There is a very nice aviation memorabilia display in the terminal lobby, including a WWII Norden bombsight! Enrique's is top rate and Ponca City is a first choice for a midday fuel stop when we are on CAP missions. (405-762-5507)

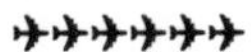

Best Mexican food in the OK and KS regions. It is hard to find good Mexican food north of the Red River — especially for us Texans. I have eaten there twice and the food is inexpensive and outstanding, especially the chips and flour tortilla. The service is quick.

Ponca City Aviation Booster Club's 

You are welcome to attend the Ponca City Aviation Booster Club's first-Saturday'-of-the-month fly-in breakfasts. We are located on the airport at the north end. This is not just your usual pancakes and sausage fare.

AMENDMENT

I'd like to echo the comment made by James Bowen concerning the pancake breakfast on the first Saturday of each month at the Ponca City Airport. My flying buddy and I tried it for the first time this past Saturday, and it was just terrific! They serve from 7:00 AM to 10:00 AM. The food's great and the hospitality can't be beat. We got there late, just before closing, but were served with a smile with what was left — pancakes, sausage, coffee. Next time we'll be on time for the omelets!

Pryor (Mid-America Industrial - H71)

J L's BBQ 

J L's BBQ is just one mile north of the Mid-America Industrial Airport at Pryor, Oklahoma. They offer some of the best BBQ in the country. Their specialty is the rib plate, and they give you a giant proportion on your plate. The airport has a courtesy car available. I highly recommend JL's restaurant for those persons traveling northeast Oklahoma.

Dutch Pantry 

It is located in Chouteau, Oklahoma, and is about three miles south of the Pryor, Oklahoma, Airport. The Pryor airport has a courtesy car so you won't have to walk. It is an Amish restaurant, serving cafeteria-style, with two types of meats at every meal. The cost is $6.00, and you can have all of the soda refills and desserts and main courses that you want. They serve homemade breads with real homemade butter. I could make a meal of the bread. I

always gorge myself and turn into a real pig, but the food is so wonderful I can't resist. There is no tipping, and seriously, I do not know how they make a profit, but apparently it is done on a volume basis.

Stillwater (SWO)

Eskimo Joe's

This restaurant is a short drive from the airport, by means of a courtesy car, in Stillwater, OK. It is world-renowned for its clothing store, which is next door. They have some of the best hamburgers you'll find, and the cheese fries are out of this world. According to their menu, the cheese fries are endorsed by former-President Bush. I've been there many times, and it's one of my favorite places to go.

Tahlequah (Tahlequah Municipal Airport - H73)

K.C. Harris Burgers

If you can remember how good hamburgers tasted when you were younger, this is for you. Kriss Harris uses secret old family recipes to create the very best hamburgers and brisket sandwiches, old fashioned ham & bean soup, outstanding cheese broccoli soup, out-of-this-world genuine north-eastern Oklahoma ranch-style chili and a dozen other indescribably delicious dishes. She uses only the freshest and finest ingredients available. Everything is prepared on the spot before your very eyes. She's there cooking Monday - Friday 1100-1900 hours, Saturday 1100-1500 hours local. Try it once and you'll be back for more. (200 E. Downing St.; 918-456-7111)

At the airport Dutch Wilhelm will provide you with the airport courtesy car and any other service you might need, including 100LL. Airport phone 918-456-8731.

Vinita (Vinita Muni. - H04)

McDonald's

This is the largest McDonald's in the world (29,135 sq. ft.) and is on the second floor of an arch, what else, that straddles the freeway south of Vinita.

At the airport, you park northeast of the runway and walk a couple hundred feet to McDonald's. There is a large souvenir shop and tourist information available. The menu? You've got to be kid-ding!

AMENDMENT

The highway that the McDonald's straddles is purported to be none other than Route 66. One of the cooler $100 hamburgers I EVER have had.

Wagoner (Sequoyah State Park - 7F4)

Sequoyah State Park Lodge

Here's one that's a winner. They have a 3,300-foot runway located in the woods. Call over the phone located in the parking area and a State Park van picks you up for the short drive to the Resort and Lodge. I had the breakfast buffet. At $4.99 total, it's a DEAL! Solid four burgers. No fuels avail-able, and watch for deer. Bet it's beautiful in the fall.

Whitehorse Cove (Private)

Whitehorse Cove Restaurant 

Whitehorse Cove has a private grass strip located 36.00.16N 095.15.64W or 5 nm and 350 degrees from 7F4. We arrived early at 0900 but were still beaten to breakfast at the Whitehorse Cove Restaurant by at least a dozen planes. This is a great stop! The restaurant is a short walk from the strip and is on a floating building right on the lake. We were seated immediately, even though the place was pretty well filled with other pilots and fishermen. The breakfast buffet is only $4.99 and offers the usual fare. I rate it a solid three burgers. During high water, call ahead. We were told that deer take over the runway trying to stay dry.

Albany (Albany Regional - S12)

Burgundy's Eating and Gathering Place 

Burgandy's truly caters to the flying public. After landing, taxi south right into Burgandy's parking lot. Half is reserved for planes, the other for cars. They offer a variety, including omelets, soups and salads, sandwiches and, of course, hamburgers. They also offer a breakfast/lunch buffet on Sunday and holidays. Prices are reasonable. Open 7:00 AM-9:00 PM daily.

AMENDMENT

Burgundy's, the restaurant at the south end of Albany, OR (S12), now also has a Thai menu in addition to the traditional American fare. It's worth trying!

Bend (Sunriver - S21)

Sunriver Lodge 

One of my favorite eating destinations is Sunriver, a popular up-scale resort community. Good food, reasonable prices, and you will be treated like royalty. Free van service from the airport to the lodge/restaurant.

The Trout House 

Another restaurant to try at Sunriver is the Trout House, which is northwest of the airport. The Lodge is due east. It's about a ¼-mile walk along the driveway leading northwest from the FBO building. The Trout House is immediately next to the marina along the river. The restaurant offers a great view of the river. You can rent a raft or canoe for a trip down the river. There is a free pickup so you don't have to row back up river.

AMENDMENT

Here's the rest of the story about Sunriver's $4,000 omelets. My introductory flight was a breakfast trip to Sunriver with my soon-to-be instructor, who offered to let me have the left seat if I would accompany him. One omelet and $4,000 later, I had my pilot certificate!

Burns (Wagontire - 81OR)

Airport Restaurant

The food is good, home-cooking prepared by half the population of the town, pop. 2. There is a motel right next to the restaurant and gas available from the pumps in front of the restaurant. The restaurant and gas pumps are just across the road from the runway. Just taxi across the road. The big problem is that the runway is just 2,000 feet long, dirt, at over 4,700 feet elevation, which pretty much limits it to the Super Cub crowd. I vote for this as "The Airport Restaurant Closest to the Middle of Nowhere." By the way, last time I checked, the town was for sale: motel, restaurant, airport, yup the whole place.

Chiloquin (Chiloquin State - 2S7)

Melita's

Chiloquin Airport has a diner on the field named Melita's. I've stayed at the motel and eaten at the restaurant and associated bar. Average rural Oregon fare.

Cottage Grove (Cottage Grove - 61S)

The Village Green

A taxiway leads from the airport proper to a parking area across the road from the Village Green Restaurant. The runway is a 3,200-foot challenge with lights and tall trees at both ends.

Cove (Minam Lodge - 7OR0)

Minam Lodge

Minam Lodge is adjacent to the now-closed Red's Horse Ranch. This is about 15 miles SW of Enterprise, Oregon, in a roadless, wilderness area. The strip takes some talent due to altitude and mountainous terrain, but they do have great food in a spectacular environment. Breakfast and lunch are about $10.00 each; dinner $15.00 flat rate.

The Dalles (The Dalles Municipal - DLS)

The Country Kitchen

The city of The Dalles is in Oregon, but the airport is across the Columbia River in Washington. There is a nice little cafe on the field called The Country Kitchen. Be advised, they are open seven days a week but close at 2:00 PM.

AMENDMENT

Gisela and Adrienne Sexton just took over the cafe. They feature daily homemade soup, pies, cakes and specials. Their flame-broiled burger is a definite winner.

Florence (Florence Muni. - 6S2)

Old Town 

An ideal way to spend a day or a week is to fly to Florence on the Oregon coast. Florence is located about midway between North Bend and Newport. The airport is about a mile from the old town waterfront district of town. It's an easy 20-minute walk on a nice day, or you can rent a bicycle from the FBO. A trip into old town makes for a nice day trip. There are a lot of shops and restaurants, as well as charter rides. Call the FBO at 503-997-8069 to arrange for car rental from one of the downtown car dealers. With a car, you can make an extended trip of a weekend or as long as you like. There are sand dunes nearby where you can rent dune buggies. You may want to partake of the many other outdoor recreational activities in the area, such as hiking and fishing.

Glenden Beach (Siletz Airport - S45)

Salishan Lodge

Salishan Lodge, at the Siletz Bay State Airport (S45) in Glenden Beach on the Oregon coast, serves an excellent Sunday brunch and gourmet meals at other times. You can overnight there and play a round of golf if you choose. The beach is a short walk, as is a shopping mall. It may come to more than $100 but it might be worth it. The airport is unattended, VFR only, with no services. It's a 3,000 foot x 60 foot paved runway (17/35).

Hillsboro (Starks Twin Oaks - 7S3)

Aileron Cafe 

Not really a "restaurant," but on the first Saturday of every month, EAA Chapter 105 hosts a breakfast at the chapter hangar at Starks Twin Oaks Airpark (7S3). This breakfast at the Aileron Cafe has become a popular monthly event for pilots out for some $100 blueberry pancakes. Twin Oaks Airpark is a nice rural strip out in the country south of Hillsboro, Oregon. There's plenty of good airplane watching and hangar flying. Blueberry pancakes, eggs, bacon, and grits, as well as coffee and orange juice are served from 8:00 AM - 10:30 AM.

Klamath Falls (Klamath Falls Intl. - LMT)

Satellite Restaurant 

The Satellite Restaurant is on the top floor of the terminal building and serves a full menu. We found the food excellent. You'll have a great view of the airport. Watch for a lot of Air Force types flying F-16s.

Lakeside (Lakeside State - 9S3)

Mexican Cafe 

A nice turf field within walking distance of town and five restaurants, one of which is Mexican. The lake also allows seaplane operations and has a dock available at the Lakeshore Lodge.

Madras (City/Cnty. Airport - S33)

Hoffy's Restaurant 

There are no restaurants on the field but two miles away in the city is Hoffy's Restaurant, full-service and open from 6:00 AM to 10:00 PM daily. The food is very good. Adjacent is Hoffy's Motel. They offer free pickup and delivery from and to the airport. The airport has a courtesy car available.

Medford (Medford Muni. - MFR)

The Red Baron 

A good restaurant! The view of the runways is great, if you are seated at ground level. The prices are fair for the food quality. Another thumbs up.

Newport (Newport Muni. - ONP)

Alex's Cafe 

The airport isn't very conveniently located, but you can still walk or get a cab into town. Good clam chowder is abundant, and there're neat coastal tourist attractions and lots of sea lions.

AMENDMENT

We flew down to Newport, OR, last weekend and had quite a pleasant time. Fuel is available from two places on the field, neither of which appears to be self-serve (read: after hours, you might be waiting until morning).

When in Newport, we do have to recommend Alex's Cafe run by a gentleman named Alex. The food is greasy, comes in large quantities, and is exceptionally tasty. It's open from 0630 to 1400 every day for breakfast and lunch. The restaurant is about a 100-yard walk south of the aquarium in Newport. Transportation to town is available from Yaquina Cab Company. Their phone is 541-265-9552 for calling ahead, and 541-265-9552 for calling from the airport. The phone is right next to the Pepsi machine.

Pacific City (Pacific City State - PFC)

Fast Freddy's 

Okay, at the risk of never being able to go there again, I'm letting you all know about this great place. There's a little roadside burger stand, Fast Freddy's, at the north end of the airport, but no real restaurants of note. On the other hand, this is about the closest airport I've seen to the beach proper. Park the plane, walk about three blocks, and you're on the sand. It's a great beach. Arrive early, since parking's really cramped. There's room for maybe a dozen and a half planes or so. You'll find out when you do the flight planning that the runway is really short, just under 1,900 feet. Be sure you've brushed up on your landings. It's paved, so as long as you land on the numbers, it's comfortable, but you don't want to be landing long here.

Grateful Bread 

About one block north is a restaurant, Grateful Bread. Excellent lunch! Another block north is Tidewater Restaurant & Lounge, good breakfast and reasonable rates.

The Fishes 

Cape Kawanda State Park is a one-mile walk north of the Pacific City Airport. You can walk along the beach or along the main road that passes by the airport at the north end. Beach access is about ¼ mile west of the airport, just off the aforementioned main road. Cross the river via the bridge and at the right-hand bend in the road you can go straight ahead to the beach route or turn right and follow the main road north. Another note, Cape Kawanda is where the famous dory fleet lands.

AMENDMENT

At Cape Kawanda Park there is a new restaurant. The Fishes has an excellent breakfast offering. It is in the same building that many may remember as the Sunset West. It has been nicely remodeled.

Portland (Portland-Hillsboro - HIO)

Airport Cafe 

The restaurant's good and is in an excellent location for watching airport operations. Not too far from the Malibu Grand Prix track if you're into racing mini-Formula-style cars.

Portland (Portland Intl. - PDX)

Airport Cafe 

This your basic commercial airport terminal with a fair cafe. Taxi to Flightcraft, get some fuel, and ask them to shuttle you over to the terminal. They'll pick you up after you've eaten. It's about the biggest airport around here, but you can land without paying a fee.

Portland (Portland-Mulino - 4S9)

Airport Cafe 

You will see a grass taxiway that leads to the Airport Cafe. If in doubt, ask the FBO for directions.

Portland (Portland Troutdale - TTD)

McMenniman's Edgefield

There is a fantastic lodge (bed and breakfast) not far from Portland-Troutdale TTD. It is not currently marketed to the aviation community, but it is very close to TTD. The place is McMenniman's Edgefield. It includes a microbrewery producing fantastic ales, a movie theater, lodging and dining. There are vineyards on the grounds, and many events occur throughout the year. It is beautifully restored and offers a unique atmosphere. The address is 2126 SW Halsey in Troutdale, Oregon. Phone 503-669-8610 to arrange airport pickup!

Salem (McNary Field - SLE)

Roscoe's Landing

Breakfast and lunch are very good. I've never had dinner there so I cannot comment, but the menu looks good and again, reasonable. Generally there are more drive-in customers than fly-in, but with CAVU, skies it's a lot different. II Morrow is located on the opposite (east) side of the field with taxi-up parking.

AMENDMENT

Suggest you try the dinner hour. Actually one of the best meals I've had was the chicken lemon herb (chicken picatta). Sharon had the beef tenderloin and it too was quite good. Service was good. We've eaten there twice and were very pleased. It isn't the same old place, thank God. It's nice to have such a good place in Salem. Honestly, this is a real gem of a restaurant. The chef is really first rate. Mind you not a three star restaurant, but excellent presentation and preparation. Service is great but needs a little polish, that will come with experience. Let's all support good airport business like this. My only complaint would be smoke that wafts out of the bar area. It could use some supplemental ventilation.

Scappoose Industrial Airpark (1S4)

Barnstormer Restaurant

Don't forget the fairly new Barnstormer Restaurant located adjacent to the Scappoose Industrial Airpark, 1S4. Excellent specialty sandwiches with aviation-related monikers.

AMENDMENTS

The Barnstormer B&B and Restaurant is now open 11:00 AM until 8:00 PM. They've added a dinner menu and hamburger sandwiches to the lunch menu.

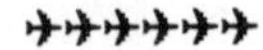

My wife and I had dinner at the Barnstormer Bed & Breakfast and found both the food and service excellent. The menu is fairly broad. I had the prime rib and my wife had the lasagna. Desserts are very good also. The Barnstormer is located just about 1/8 mile west of the airport grounds at the northwest corner. There is a small black-topped area, for aircraft parking, next to the campgrounds, near the run-up area of RWY 15. The setting is somewhat rural, and you might be able to make friends with the B&B's cat or perhaps the horses in a nearby corral.

Seaside (Seaside State - 56S)

Pacific Way Cafe 

About a 10-minute walk from the airport is a pizza place. A half-hour walk will get you into the boardwalk touristy part of town, where there's at least one good cafe. The town abuts the beach. It's an easy, about 15 minutes, taxi or bus trip to Cannon Beach another very popular coastal town.

AMENDMENT

The closest restaurant to the Seaside Airport, and one of the best on the northern Oregon coast is the Pacific Way Cafe in Gearhart. It's a closer walk from the airport than the Seaside restaurants. Walk north for a few blocks on the highway, turn West on G street, turn right at the only stop sign, and go a couple more blocks to the restaurant. It's exactly one mile by my car odometer from the airport to the restaurant.

Tillamook (Tillamook - S47)

Air Base Cafe 

Tillamook S47 in Oregon now has a restaurant in the giant blimp hangar called the Air Base Cafe. They offer the usual hot sandwiches and burgers served with chips. If you would like, for dessert they offer Tillamook Ice Cream. The restaurant is open from 9:00 AM to 4:00 PM every day. A very interesting place to visit, The Tillamook Naval Air Station Museum offers specially reduced prices for pilots who would like to visit the museum and also a discount on aviation fuel.

Allentown (Allentown - ABE)

Creative Croissants

I work at ABE right next to the restaurant at Continental Express. It has been recently remodeled and is now called Creative Croissants. They have a fairly large variety on the menu, geared to healthy foods, specializing in stuffed croissants filled with vegetables and meats.

Sheraton Jetport Inn

They have an excellent Sunday breakfast 8:00 AM - 10:00 AM and lunch 11:00 AM - 2:00 PM. The FBO will help arrange transportation to the restaurant. (610-266-1000)

Gregory's

The restaurant at ABE is known to the locals as Gregory's. They offer the best and biggest steaks you have ever seen or chewed on! You may choose from a 6 oz, 12 oz, 16 oz, 24 oz, 48 oz, 72 oz, 96 oz, and 120 oz. for the above-normal eater. Prime rib goes up to 30 oz. The best part is the pricing. A 24 oz. ribeye 1½ inches thick is $10.95, the 48 oz. is just $19.95. Prime rib prices are $19.95 for 30 oz. You can't miss them. They are located on the final approach for runway 24, about one mile from the airport. By the way, they also have the best barrel of pickles you'll ever want to taste.

Allentown (Queen City Airport - 1N9)

The Cascade

Queen City Airport (1N9) at Allentown, PA, has several passable eateries within walking distance. My favorite is The Cascade, a classic diner offering good value. Enjoy!

Altoona (Blair Cnty. - AOO)

The Kittyhawk Restaurant 

The Kittyhawk Restaurant has very good food and a friendly staff. Don't plan on eating there if you land after 8:00 PM.

Bally (Butter Valley Golf Port - 7N8)

Butter Golf Port Snack Shop 

A nice down-home place, open during the summer, is at Butter Valley Golf Port, located in Bally, PA. Bally is about 50 miles north of Philadelphia, just east of Rt. 100. In addition to being a golf port, Butter Valley has a quaint clubhouse where flying regulars get together for Sunday breakfast for about $2.85 a head. If they see you landing, they'll have your eggs ready by the time you get in. If you like golf, that's good, too!

Bradford (Bradford Regional - BFD)

Piper Pub Restaurant 

Check out the Piper Pub Restaurant on the field at Bradford, PA. Bradford has a huge but little-used runway with ILS. It is situated in the middle of the nicest countryside in the state. Check the weather at the on-field FSS, open until 4:00 PM. Don't forget to low-pass at night to check for deer.

AMENDMENT

Quiet, clean and in the terminal overlooking the ramp. Prices are fair and they offer a special each day. The food was average, but service today was not the best. We'll go back, but only if we are in the area.

Butler (Butler Cnty. Airport - BTP)

Runway Restaurant 

The Runway Restaurant has just opened at the Butler County Airport in Butler, PA. First impressions are great! We have fine facilities at our field, including ILS and VOR approaches, VASI, AWOS, remote clearance frequency, etc.

AMENDMENT

Definitely better than most airport restaurants and conveniently located at the north end of the airport right by the ramp. I'd give the Runway Restaurant four burgers. During a Sunday brunch the food was better than average, although there wasn't the variety that you'll find down at LBE's restaurant. Reasonably priced at $8.95 for the Sunday buffet, the only downside to the Runway is that it's popular enough with the locals that you might end up waiting 20 minutes before you can be seated. If you're willing to sit at the bar though, odds are you'll be served immediately.

Coatsville (Chester Cnty. - 40N)

The Flying Machine 

A really nice place in the terminal building at 40N. The FBO is based in the same building, so it is easy to order fuel on the way to a burger.

AMENDMENT

I just visited Chester County...great food in the main building, large landing strip, even for commercial jets. It is a great place to fuel up and eat on your way north or south, away from Class C and B airspace of Philly and Baltimore.

DuBois (Jefferson Cnty. - DUJ)

The Aviator's Rest 

The Aviator's Rest, a luncheonette-style restaurant on the field is decorated with WWII memorabilia. I loved the hot wings!

Easton (N43)

Fork's Valley Inn and Restaurant 

Fork's Valley Inn and Restaurant is only 1/4 mile from the airport. There is a courtesy car available at the FBO, but you won't need it.

Franklin (Venango Reg - FKL)

Primo Barone's Restaurant 

Primo Barone's Restaurant is located in the modern terminal building. On my first visit I was very impressed. The restaurant was close to full on a Sunday afternoon. The menu is mostly Italian, but with a good sandwich selection. Very clean and decorated with aviation items, it provides good view of the runways. Service was quick and the prices were reasonable. I will definitely fly in again.

The restaurant is located right on the side of the ramp in the terminal. Great view of the runway and tie-down area. The food is great. I would try the N.Y. strip steak which is 1-inch thick! The service is quick and friendly. The prices are good and reasonable. The setting is good, with dimmed lighting. Great for a romantic Italian dinner.

Gettysburg (Doersom - W05)

Herr Tavern 

Historic and delicious dining in Gettysburg at the Herr Tavern.

For the perfect weekday and Saturday lunch stop, I recommend the Herr Tavern just outside Gettysburg, PA. Land at Doersom (W05) and enjoy the pleasant ¼-mile walk down Chambersburg Pike. Herr Tavern has wonderful sandwiches as well as full-course meals and daily specials. The Tavern also offers B&B accommodations that are first class. The Herr Tavern's phone is 717-334-4332. By the way, there is a $5.00 landing fee at Mr. Doersom's Airport, but well it's worth it for the convenience and hospitality of this little airport. Overall, I would rate the experience AT LEAST "five flying burgers." Enjoy!

AMENDMENT

We went to Doersom and the Herr Tavern. It is one of the best places I have flown for an enjoyable lunch.

Harrisburg (Capital City Airport - CXY)

Coakley's Irish Pub

Looking for another new place to eat in central PA? Try Coakley's Irish Pub in New Cumberland, PA, by the Capital City Airport. Coakley's is a renowned Irish pub serving lunch and dinner in a great atmosphere. It is located a block or so from the airport. If you taxi up and park at the Harrisburg Jet Center, the line boys are always glad to give you a ride.

Pierre's Steak House

Pierre's is know locally for serving excellent steaks since 1929. Like Coakley's Irish Pub, it is located a block or so from the airport. If you taxi up and park at the Harrisburg Jet Center, the line boys are always glad to give you a ride to either restaurant.

Johnstown (Johnstown-Cambia Cnty. - JST)

Skyway Lounge

The airport at Johnstown, PA, has a great little restaurant. Their specialty is Italian.

Kutztown (Kutztown Aviation - N31)

Airport Diner

The Airport Diner at the north end of the field is one of the best diners in the area. Its right on the field and is open 24 hours a day.

Lancaster (Lancaster Cnty. - LNS)

Lancaster Airport Restaurant

Need some work at a towered airport and want a great meal to boot? Try the restaurant at the terminal building at the Lancaster Airport (LNS). The traffic controllers in the tower and the waitresses will treat you like family. On a beautiful weekend afternoon, the airport gets its fair share of traffic, but the food at the restaurant makes it worth the effort. Sample the Pennsylvania Dutch-style cooking and have a great flight as well.

AMENDMENT

We stopped at Lancaster for lunch last Sunday and ended up taking a 15-minute walk to the local grocery store/nursery. Great prices on a variety of plants and lots of craft stuff, too. They are located out the main entrance to the right then right, again to the stop light.

Latrobe (Westmoreland County - LBE)

Blue Angels

Blue Angels is in Latrobe, PA, Arnold Palmer's home town! Great golfing in the area, obviously. The restaurant features great Italian as well as good hamburgers.

AMENDMENT

Their food is excellent! In particular, they have a Sunday morning all-you-can-eat brunch for about $9.00 per person. You can make one pass through the buffet line to get breakfast (I particularly like the omelets); a second pass through for lunch, which features steamship roast beef, ham and turkey; and a third pass for dessert. Well worth the flight.

Monongahela (Rostraver - P53)

Eagle's Landing

I usually go to Eagle's Landing at Rostraver (P53) for my $100 hamburger. It has a nice atmosphere and average food.

Philadelphia (NE Philadelphia - PNE)

94th Aero Squadron

This is one of a chain of restaurants that are almost always built on or adjacent to the airfield, usually with direct airport access. Views of the runway are typical. They have an excellent menu and tend to have moderate to higher prices. This chain is several steps above your typical $100 hamburger. The motif is a WWII bombed out English farm building with various aviation artifacts.

Pottstown (Pottstown Muni. - N47)

Pottstown Diner

If you find yourself in the pattern at Pottstown Muni (N47) on downwind for 07, look about 1,000 feet to your left just before you're abeam the numbers. That big road going by with the strip mall just there is Rt. 100, home of the Pottstown Diner! If you can't hitch a ride from someone and you're on foot starting from the FBO, just head out the gate and straight ahead. It is one LONG block out to Rt. 100. It's at the southerly end of that strip mall right in front of you. I've always been in a car, which makes it a one or two-minute drive, but I'd guess it to be a ten-minute walk or so.

This is a genuine diner in the tradition of the famous Pennsylvania diners. It's one of the later ones that's been expanded over the years rather than the trailer style. The atmosphere is comfortable and family-oriented.

There's a very nice soup and salad bar that shouldn't be missed and lots of home-style items to order from the menu. The prices are reasonable, above $5.00 but less than $10.00 for just about any meal. It's a nice friendly place! I'd recommend it...say...I just did!

Reading (Reading Regional - RDG)

Wild Wing Cafe 

Try the Wild Wing Cafe at the Reading Regional Airport (RDG). Excellent buffalo wings, burgers, etc. Two large (5,000+ foot) runways and radar approach facility make this easy to get to. Ask for a window seat and watch the bizjets take off and land. Also, there is great shopping at the local factory outlets if you have some time. Wild Wings Cafe reservations are at 610-478-1747.

AMENDMENT

The Wild Wings Cafe is a popular spot among the locals, so there is usually a crowd at the restaurant in the evening. If you like HOT wings, this is the place to go. The "suicide wings" are extremely hot; try one before you ask for more. You may want to have a designated pilot who doesn't eat them. There is limited parking north of the restaurant. The spaces aren't marked very well, so you just have to fit in.

Seven Springs Borough (Seven Springs - 7SP)

Seven Springs Resort 

For something a little different from the usual airport burger, try a fly-in Sunday brunch at Seven Springs Resort, PA. The runway (10-28) is 3,045 feet x 42 feet and is located on top of a ridgeline. After tying down, use the phone at the little airport chalet to call for the courtesy bus that will take you down to the main lodge. The brunch is very lavish, with salads, fresh fruits, roast beef, crab legs, quiche Lorraine, Belgian waffles, made-to-order omelets and desserts galore. The price is $14.95. By the way, the Lodge is very "kid friendly," if you are traveling with young aviators. We had to do a new CG calculation after this feast!

Smoketown (Smoketown - 37PA)

T. Burkes 

This is one of the best restaurants I've been to. Its just a few hundred yards walk, mere minutes, from the Smoketown, PA airport in the heart of PA Dutch country. It's called T. Burkes and is referred to locally as the "deli." It's a little more upscale than the typical airport greasy spoon. The pleasant atmosphere, combined with good food and reasonable prices make this a must-stop if you want to impress your passengers. You're very likely to encounter Amish buggy traffic near this airport.

Craig's Lunchbox 

Craig's Lunchbox is a cozy sandwich and soup shop located just behind the FBO office at Smoketown Airport. They are open Monday through Saturday. Saturday hours are slightly shortened to 8:00 AM to 2:30 PM. The sandwiches are excellent and inexpensive.

Tunkhannock (Skyhaven - 76N)

Cross Country

Great place to fly in. Next door to the field is the Cross Country restaurant, which serves great home-style meals, especially on weekends. The strip is small, 1,900 feet, and has been recently paved.

Waynesburg (Greene Cnty. Airport - WAY)

A.J.'s Landing

Greene County (WAY) has a new restaurant. The restaurant is A.J.'s Landing.

Wilkes-Barre (Wyoming Valley - WBW)

Victory Pig Bar-B-Q

Here's a place to get something a little unusual. If you fly into Wilkes-Barre/Wyoming Valley Airport (WBW), you can take a short walk, less than a 1/2 mile north on Route 11, to the Victory Pig Bar-B-Q, next door to the place with the golden arches. Their pizza is famous thoughout Wyoming Valley, and people drive for miles to get it. They are only open Wednesday, Friday, and Saturday nights.

You can also get in a game of miniature golf or smack a few on Bob's driving range, which is located next door.

Colonial Pancake House

Colonial Pancake House is a short walk across the street from the ramp. Looking for a buffalo burger? Try this place. Most pilots stop here for coffee. There are no fees for short parking at the airport.

Williamsport (Williamsport Airport - IPT)

Skyview Restaurant and Lounge

The Skyview Restaurant and Lounge is on the second floor of the terminal building. The giant picture windows give a great view of the airport action. This is my home field. Come visit!

AMENDMENTS

Good place! If you buy fuel at the FBO, they'll give you a card for a 10 percent discount at the restaurant.

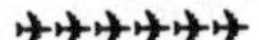

I had to fly into IPT from Toronto late one night. We ended up at the Skyview. What a great atmosphere in the evening! I actually had the BEST LOBSTER I've ever eaten, at an unbelievable price! Try this place!

York (York - THV)

Orville's Restaurant 

Orville's is a restaurant enjoying increasing popularity with flyers and locals. Don't miss The Flyer's Den, the best pilot shop I've visited in the region.

AMENDMENTS

Orville's Restaurant at York (Thomasville - THV) Airport in south-central Pennsylvania has very good.....cheap.....food served in a nice atmosphere.

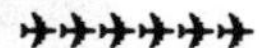

Orville's has recently doubled their seating capacity, so no more waiting! Always the very freshest ingredients and imaginative menu specials. Terrific breakfasts, homemade soups, good salads, and a nice list of dinner entrees. You would expect to pay more for food this good, but the prices are quite reasonable. This is the very best fly-in restaurant I have discovered.

Block Island (Block Island State - BID)

Earnie's

Located right at the airport. They waive the $5.00 state landing fee if you only go to the restaurant. Open Memorial Day through Columbus Day. There are many restaurants located in town, a 20-minute, 1-mile beautiful walk or a $5.00 taxi ride. They vary in price, but most feature good sea-food and are located right on the ocean. Block Island is well worth a day or weekend visit!

AMENDMENT

If you get there before 11:00 AM, make certain you catch breakfast at Earnie's!

Providence (Green State - PVD)

Hooters

N2679H and I like to fly to Providence, RI. Hooters is a five-minute walk from the FBO. Keep on Coupin.'

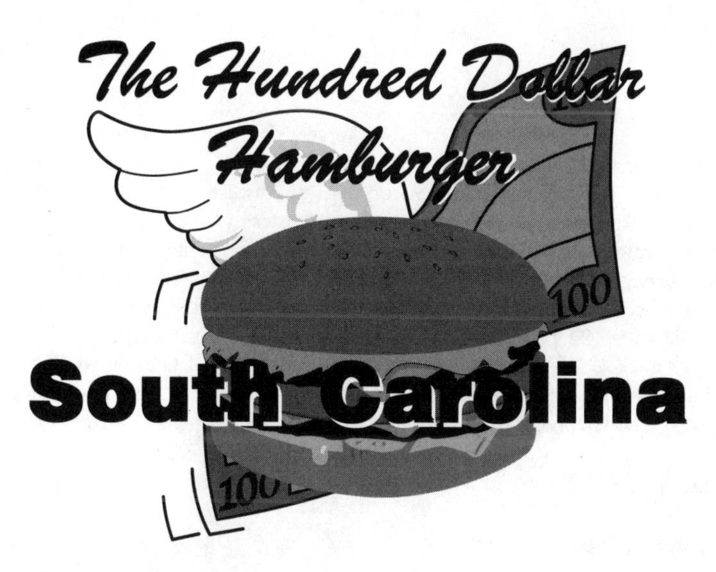

Bamberg (Bamberg Cnty. - 99N)

Tommy Rose Barbecue 

Many of us have visited Tommy Rose Barbecue in Bamberg, SC (99N). The restaurant is within walking distance, but the owner is happy to come to the airport and pick you up, then deliver you back after your meal. The food is excellent, cheap, varied, and the service is great.

Barnwell (Barnwell Cnty. - BNL)

Winton Inn Restaurant 

I have found a little-known place in South Carolina. After flying into Barnwell Co. Airport, the Winton Inn Restaurant will send a car to pick you up. A short drive later and you are enjoying an all-you-can-eat seafood meal, including Alaskan king crab legs. The meal is $17.00 per person. There's not a whole lot of atmosphere, but the food is great.

Charleston (Charleston Exec. - JZI)

Cappy's Seafood 

The airport is actually on John's Island, and there is no place to eat at the airport itself. A short ride away is Cappy's Seafood. They have a great selection of fresh seafood in a nice atmosphere. I was actually expecting it to be pricey, but it turned out to be quite cheap. Go on a nice VFR day and sit outside.

Gibson, NC (Stanton's BBQ - Private)

Stanton's BBQ 

Stanton's BBQ is a real fly-in restaurant. About 20 steps from the plane parking is the large red restaurant building. It's a private field, not on the sectional. The coordinates are N 34.43.9 W 79.40.0. If you're flying from the Florence, SC, to Sandhills, NC,, VOR route, you'll fly very close. It's near the town of Gibson on the NC/SC border, southwest of Fayetteville. The restaurant is immediately adjacent to the field. Be sure to tell the hostess you're a pilot. They've got a special room waiting for you that is for pilots/families only. The strip is smooth grass — about 2,200 feet long — easy, clear approaches. They're open every day for lunch, including weekends. (The strip is not lighted.) The fare is basic country cooking — the BBQ is great. This is a great place to take the family. The Unicom is 122.9. Call for "BBQ traffic" and you'll get a friendly response. Stanton's BBQ is actually in SOUTH CAROLINA, - but three miles SW of Gibson, NORTH CAROLINA. Since they don't put that nice state line on the sectional, most of us truly don't know which state we're in anyway! The telephone is 803-265-4855.

Hilton Head (HXD)

Pizza Hut

A short walk down the road leads to a sort of shopping center. It has a Sam's Club, along with several restaurants. I ate at the Pizza Hut. I believe there was also a McDonald's and a Taco Bell. Short on character, long on convenience.

Myrtle Beach (Myrtle Beach Jetport - MYR)

Landmark Best Western 

I fly to Myrtle Beach Jetport (MYR) often, and if you like South Carolina beaches, it can be a lot of fun. I stay at the Landmark Best Western, 803-448-9441. They will pick you up at the airport, upon request. The hotel is on the beach and has an indoor/outdoor heated pool. Also, the town has frequent bus shuttle service to other beachfront shopping areas and amusement parks. The public buses stop right in front of the hotel. I also understand that they have some great golf courses. Room prices run somewhere between $50.00 to $75.00 per night, but since you don't need a rental car, I feel it's a bargain. The only snag is that the airport has just recently converted from a joint-use military/commercial airport to civilian operation. I'm not sure if a fixed-base operator has started operating on the field yet. North Grand Strand Airport is also open to private aircraft, but I've not been there and don't know how far away from the beaches it is or if transportation is available.

AMENDMENT

Here's the FBO info: Myrtle Beach Aviation (Jetport) 1-803-477-1860, 1-800-474-KMYR, Fax 803-477-1769, 2551 Phillis Blvd., Myrtle Beach, SC.

North Myrtle Beach (Grand Strand Airport – CRE)

River City Cafe

Grand Strand Airport (CRE), North Myrtle Beach, SC, is my favorite destination when I'm looking for someplace to go.

CRE has one paved runway (5-23), VOR on the field (CRE), tower (nonradar) and two FBOs. I strongly recommend Ramp 66 for services. A really sharp line crew will meet you as you taxi in and help you park. When it's busy, and it usually is in summer, they will park you at the door and then tow your plane to tie-down. They have a courtesy car available for free for one hour. If you will be longer, you can rent by the hour or day. The car rental is $8.00 per hour, with the first hour FREE. This keeps down abuse of the courtesy car!

Many, MANY places to eat within a couple-mile drive. A couple are within walking distance. The beach is about a ¾-mile walk down a residential street. Watch out, a couple of people on the street have dogs that don't appear to be chained, but I've never had one come after me. They just bark a lot!

The main area of Myrtle Beach is about a 5-10 minute drive away, one of the biggest tourist traps you'll ever see! They have EVERYTHING!

Also, they have an extensive banner tow operation. I can sit for hours and watch them drop and pick up the banners. Those guys are great. It gives you a whole new perspective on maximum performance operations watching a C-150 pick up a 60-foot banner from the ground!

AMENDMENT

For "the best burgers on the beach" try the River City Cafe, located at 21st Ave North in Myrtle Beach. This is about 15 minutes by car from CRE. Head south on Rt. 17 and stay left on Business 17 (NOT Bypass 17) until you get to 21st Street. Then turn left and you're there. They feature salted in-the-shell peanuts. Be sure to toss the shells on the floor. They have numerous varieties of burgers and sandwiches. Their bloomin' onion appetizer is an absolute must, and their onion rings are superb.

Horry's Seafood 

I often fly to North Myrtle Beach during the fall and winter for a visit to Horry's Seafood. They have the best oyster roaster around. If you don't like oysters, the fried fish is outstanding, and of course you must try the onion rings and corn dodgers. Horry's is a simple restaurant, more like an old fish camp. The atmosphere is friendly and the food is great. Horry's is located north of Strand Airport (CRE) about 8 or 9 miles. Take Hwy. 17 north out of the airport to Hwy. 9 north. Horry's is located on the bank of the Waccamaw River on your right. The folks at Ramp 66 will work a deal for you on a courtesy car or a cheap hourly rental.

Spartanburg (Spartanburg Downtown Memorial - SPA)

Beacon Drive-In Restaurant 

Spartanburg, SC, is the location of the world's best pork barbecue at the Beacon Drive-In, located north of the airport about two miles on Reidville Road. Just ask anyone at the field. Try the "outside sliced with extra sauce," a load of fried onion rings, and iced tea and notify your cardiologist. Heaven...

Wall (Wall Muni. – GV4)

World Famous Wall Drug 

A wonderful place to fly-in for breakfast is Wall, SD (GV4). The airport is about five blocks from the center of town and the World Famous Wall Drug. For the uninitiated, Wall Drug is a unique combination of shops, restaurants, and tourist attractions that got its start by advertising "Free Ice Water." Food is very reasonable in price, and you can eat in one of several dining rooms, all decorated with wonderful collections of Western art. The town also has several other good eating places (Cactus Cafe, Elkton House) that can give you a change of pace from the good food at Wall Drug.

The airport is infrequently attended. There is a pay phone where you can call for gas. Oftentimes, people see you land and insist on driving you the few blocks into town.

It's a fun change of pace.

Chattanooga (Lovell Field - CHA)

Rib and Loin 

Go to Krystal South and ask one of the lovely ladies behind the counter if you can use their car. Buy a little gas, not a lot, as it's way over $2.00 a gallon. Leave the airport and make a right. At the next light make a right and you will be on a main drag. Go about a mile or so and it will be on the left. It's really good barbeque.

Homer's

On the other side of the street from the Rib and Loin is an all-you-can-eat place called Homer's. For $5.50 per person it's not too bad. Just don't buy any seafood. Stick to steak!

Dyersburg (Dyersburg Muni. - DYR)

Mary Lou's Restaurant 

They serve plate lunches and burgers for less than $100! Everything is very good!

Elizabethton (0A9)

Betty's Burger Bar 

Betty's Burger Bar is right across the street. Good food! Super milk shakes!

Greeneville (Muni. - GCY)

The Stockyard 

Borrow a car from the FBO and eat at the Stockyard. It is open Monday through Saturday from 5:30 AM until 3:00 PM. Great food!

Knoxville (McGhee-Tyson - TYS)

The Belair Grill

The Belair Grill is near McGhee-Tyson, Knoxville, TN (TYS). The absolutely best hamburgers anywhere! Park at Cherokee Aviation and ask for the courtesy car and directions. If you borrow the car, please buy some of their gas.

P.J.'s Landing

I am based at Knoxville TYS. The Belair Grill is good for a nice sit-down meal. If you want a different kind of adventure, P.J.'s Landing is a marina not five minutes from the airport. They have a decent little deli at the dock. Anyone at Cherokee Aviation should be able to give directions. It is on Louisville Road, just past the Texaco station. Take Hunt Road, just past the baggage terminal, to Louisville Road.

MacKinnon (Houston County Airport - M93)

Southernair Hotel and Restaurant

There is a very cool restaurant called the Southernair Hotel and Restaurant near the Houston County Airport (M93). The landing strip is paved and 3,700 feet long. No services are available at the airport. Park your plane and walk 300 yards to some of the best Tennessee catfish in the world. Houston County is situated just north of the I-40 Tennessee River intersection and south of Fort Campbell MOA.

McMinnville (Warren County Memorial - RNC)

Country Ham FlyIn

Warren County Memorial (RNC) has a fly-in country ham breakfast on the second Saturday of each month. Call Joe Howard at 615-668-4806 for info. Good ham and classic planes.

Memphis (General Dewit Spain - M01)

The Rendezvous

Flying into Memphis, you can smell the ribs just as you have to call for clearance into the "bravo" airspace. Skip Memphis International. Fly just north of downtown to a small 3,800 foot strip locally know as Spain. General Dewit Spain (M01) is primarily an agricultural airport. Talk nice and they may lend you a car. If not, take a cab. The Rendezvous is situated in a basement accessed through an alley directly behind the Peabody Hotel. Ask a local for directions. They will know. Order a full rack of ribs and get ready. The waiters are rude, the place is hard to find, and the beer is cold. You may want to stay the evening at the Peabody or have a designated flyer. The ribs are absolutely the best on earth.

Nashville (Nashville Intl. - BNA)

101 Airborne

The 101 Airborne is a WWII style French country home located on the southwest corner of the airport, overlooking 21L/21C. The food is great. Although this is better than standard airport fare, the prices are reasonable. Stop at Steven's Aviation instead of Signature and you'll save enough on that stupid service charge to pay for dinner. Lots of WWII goodies, including old photos, newspapers, and big band music.

NOTAM: Watch for bombed-out bridge and tank on approach.

Nashville (John C. Tune Airport - JWN)

Blue Moon Cafe

Fly into John C. Tune Airport (JWN) and take the courtesy car to the Blue Moon Cafe. Local flavor. Catfish and seafood. Sometimes you may catch a country music star or some unknown playing music. Great place to eat and only two or three miles from the airport. Ask someone at the FBO for directions.

Smyrna (Smyrna Muni. - MQY)

Bill's Bar and Grill

Bill's is located across the street from the Smyrna Air Center and offers the usual grill items.

Tiptonville (Reelfoot Lake - 0M2)

Reelfoot Airpark Inn

Reelfoot Lake State Park in Tennessee is an excellent place to visit. Reelfoot Lake was formed back in the late 1800s as a result of a very strong earthquake along the New Madrid fault. There is a restaurant about 125 yards from the airstrip. It serves a very tasty buffet-style lunch. Camping facilities, condo-type rooms, guided fishing trips and sightseeing trips all make this a great stop. I do not know anything about the quality of the facilities other than the restaurant, but I rate it very high on my list of places to buy a $100 hamburger. It stays open from about the middle of March until sometime in October, if memory serves me correctly.

AMENDMENT

I agree, Reelfoot is a favorite. The lake is famous for the number of eagles, ducks and geese that can be seen. Eagle tours are offered at 8:00 AM and 10:00 AM for $3.00 per person. The tour bus makes frequent stops at state supplied telescopes. I normally see 5-15 eagles. They migrate out in March. Take the birds-in-the-vicinity warning seriously. Flocks of 500 ducks or geese are not unusual.

Tullahoma (Tullahoma Reg. - THA)

Piggy's Place

Tullahoma, Tennessee, is a fine destination for good food, the Staggerwing Museum and Jack Daniel's Distillery in nearby Lynchburg, Tennessee. The FBO has two courtesy cars available for fuel-buying customers. Piggy's Place has unusually good barbecue.

Abilene (Abilene, Regional - ABI)

Joe Allen's BBQ 

Get the loaner from Abilene Aero and go downtown to Joe Allen's BBQ. The FBO has maps available. It's some of the best Texas BBQ anywhere. Just be sure you land at the eastern airfield, not the western one.

AMENDMENT

Joe Allen's B-B-Q is so good! Out-and-back sorties during USAF pilot training were flown to ABI just to go eat at Joe Allen's.

Amarillo (Amarillo Intl. - AMA)

The Country Barn 

The Country Barn has to the best place to eat in Amarillo! They have excellent chicken-fried steaks! It's close to the airport (AMA), and the folks at TAC AIR will give you a ride, then come pick you up when you're ready. The lavatory basins are shaped like the state of Texas. TAC AIR is a full-service facility and is always open. This pilot has to give The Country Barn "two thumbs up."

Andrews (Andrews Cnty. - E11)

Buddy's Drive Inn 

If you like REAL steak fingers go to Buddy's Drive Inn. The steak fingers are sliced from round steak fillets and deep fried to a golden brown. The portions are huge; you WON'T go away hungry. The price is about $6.00 per person, and they have doggie bags on hand. The Andrews County Airport (E11) is about seven or eight blocks east of the cafe. There is a real nice Ford Crown Victoria courtesy car waiting for your arrival. The airport has three runways, two lighted. Buddy's deserves at least a four hamburger rating.

AMENDMENT

Anyone can say what they want about anywhere else having the best fried steak, but Buddy's beats them all. Once you've experienced it, you'll be back for more!

Arlington (Arlington Muni - F54)

Wings

I am somewhat of an airport cafe freak, having braved a crossing of the mighty Pacific Ocean, well, sort of, just to get my oil-stained paws on one of those famous buffalo burgers at the Runway Cafe on Catalina Island, CA. There's nothin' like sitting out on the patio tasting those wonderful patties all the while watching my meal's brothers and sisters roam on the range.

Well, the Wings Cafe doesn't quite have that atmosphere, but it does have VERY good hamburgers and fries. I fly out of Arlington and eat there about once a month and have never been disappointed. The menu is what you might expect from this kind of café, with the typical sandwich names like the Stearman Burger, etc.

Prices are certainly reasonable, so much so you just might want to take your instructor there for lunch after your biannual. I did, and it worked!.The decor used to be airplane memorabilia. It seems the owners are changing the program a bit to fit the sports nuts who like to spend money here in "Ranger land."

The Cafe is about midfield in the ARLINGTON JET hangar on the second floor. It faces the runway, so on a good Sunday you can eat and watch those poor little C-152s bounce down the runway to your heart's content. It is a somewhat small place, and I would guess it seats about 60.

Don't know what the hours are, but they must be reasonable as I am rarely disappointed, and the coffee is always hot and STRONG.

AMENDMENTS

Wings is open 8:00 AM - 5:00 PM Monday through Friday, 9:00 AM - 4:00 PM on Saturday and closed on Sunday.

Just taxi up to the door at Arlington Jet Center and the line personnel will take care of your ship. Upstairs, the Wings cafe serves breakfast and lunch. Place your order, with the cashier, from the menu on the wall and take a seat. Service is good and prices are low. I ordered the club sandwich with chips and a coke for $5.00. I have seen bigger clubs before, but it was fresh and tasty. You can pick up some Cessna parts, as there is a big dealership on the field.

Austin (Lakeway - 3R9)

Lakeway Resort

Lakeway Resort on Lake Travis has one of the best Sunday buffets in Texas. Land at the airport and call the inn. They will send a van at no charge. Feast while you look over Lake Travis and the hills of Texas. True four-star dining!

Beaumont/Port Arthur (Jefferson Cnty. - BPT)

Dorothy's Front Porch

About two miles from the Jefferson Co. Airport (BPT) is a great home-cooking eatery called Dorothy's Front Porch. While it's not right at the airport, the nice guys at the FBO can be coaxed into giving you a ride there and back. Dorothy's Front Porch is built over a large lake. The lake is populated by dozens of ducks. Free "critter food" is provided for feeding them. The restaurant itself specializes in catfish, with the smoked variety highly recommended. You can escape for about $8.00 - $10.00 a head. Dessert is a necessity. All of the cakes, pies, etc., are homemade. The "Slice of Heaven" is worth the trip. Dorothy's Front Porch is open every day except Monday, until 9:00 PM.

Bellville (Grawunder Muni. - 06R)

The Bellville Cafe

Bellville, Texas (06R), is a great place to fly to when you are just looking for an excuse to aviate. The airport is located right on the edge of town. A short walk gets you to the old Bellville town square, which is having a rebirth as an antique center. There are numerous good restaurants on or near the square. The Bellville Cafe has the definitive CFS, that's chicken fried steak — the national food of Texas for you foreigners.

My wife and I have spent the better part of a Saturday touring Bellville on foot and thoroughly enjoyed the place. Bellville is located in the heart of Bluebonnet country and is really scenic in the spring.

The airport has a single runway, 17-35, about 2,400 feet long, sloping uphill, with the 17 approach end about 45 feet higher than the far end. Landings and takeoffs are not difficult, but they certainly aren't routine!

Brownsville (South Padre Intl. - BRO)

Louie's Backyard

The BRO restaurant is located in the airline terminal building, with picture windows overlooking the field. It specializes in Mexican food, hamburgers and french fries, but all other items are also available on the menu. Prices are very reasonable for an airport restaurant. It is operated under contract by the well-known South Padre Island operator, Louie's Backyard. Louie's Lounge is adjacent to the restaurant.

As the field is air carrier rated, access is not available directly from the ramp but only through FBOs, which are located conveniently on each side of the terminal building.

Barnstormer's

The CAF has a wing here, and in the corner of their hangar is Barnstormer's, a great bar with about a billion airplane models hanging from the ceiling and plenty of stories to go around. The CAF hangar is located south of the tower. Nearby, you can see corpses of a PBY Catalina, a DC-3, and a Viscount that was parked flyable 15 years ago but has sadly deteriorated. Not much food to be had at this bar, but the atmosphere is the best.

Cameron (Cameron Muni. Airpark - T35)

Big Bob's Steakhouse 

Cameron Muni is a well-maintained hard-surface strip two miles north of Cameron, Texas. The runway is 3,200 feet x 50 feet and well lighted. Big Bob's Steakhouse is a small friendly place located about 3/4 of a mile from the field on Industrial Blvd. It is very clean, with a truly country flair. The hospitality is gracious and the steak out of this world. The prices are reasonable and the portions generous. The cheesecake is homemade, so be sure not to miss it. Believe me, after such a big dinner the short walk back to the field is a welcome relief. Should you prefer not to walk, just call ahead and someone will shuttle you to and from. Open Tue. – Sat. 6:00 PM to 10:00 PM. Phone 817-697-4669.

Canton (Canton-Hackney Field - 7F5)

Ranchero Restaurant 

The airport at Canton, Texas, famous for First Monday Trade Days, is a nice place to go for breakfast, lunch, or dinner. About a mile east of the field is a wonderful buffet at the Ranchero Restaurant at a very reasonable price, $6.95 for all you can eat. It's next door to a What-a-Burger! The Canton First Monday Trade Days is a huge flea market. It covers several acres and takes place the first weekend of every month. People, especially in east Texas, are very familiar with it. It has been going on for years.

Cleburne (Cleburne Airport - F18)

Mary's Home Cooking 

This is not your usual airport restaurant. Mary has been cooking Louisiana and TexMex style for the last 15 years. She was awarded a five-star rating by the Galloping Gourmet who visited her New Orleans restaurant! She is also quite good at meatballs and pasta, barbecue and TexMex cornbread. Yes, you can get a hamburger, but what a waste when so many other goodies await you! Mary's is open for lunch and dinner Monday through Friday and breakfast on Saturday, starting at 0700. Mary's is in the big hangar on the south end. You can taxi directly to her door. (817-517-6630)

AMENDMENT

The hours of operation have changed, Monday-Wednesday 11:00 AM to 5:00 PM, Thursday-Saturday 11:00 AM to 8:00 PM and closed Sunday. The food is predominately TexMex. It's just a friendly and enjoyable place to fly in and have a meal.

College Station (Easterwood Field - CLL)

Tom's Bar-B-Que & Steakhouse 

There's a really terrific barbecue and chicken-fried-steak emporium in College Station, Texas. Tom's Bar-B-Que & Steakhouse on South Texas Ave. It's located near the Texas A&M campus and draws not only college and local family business but a great deal of fly-in traffic.

Their barbecue, whether you choose beef brisket, ribs, sausage, pork or chicken, is always tasty.

They offer a sampler plate, so you can choose two or even three different meats. Many university students get the Aggie Special, a big slab of brisket or sausage served on butcher paper with a hunk of cheese, a big piece of raw onion, a dill pickle and several slices of light bread, as they say in east Texas. Your utensil for getting this repast into manageable pieces is a large butcher knife. The accompaniments, beans, potato salad, slaw, iced tea and so on, are a clear cut above average. The chicken-fried steak equals the barbecue in quality and quantity of the serving. This is my favorite barbecue joint. Can you tell?

Fly into Easterwood Field (CLL), a well-maintained, friendly facility, and borrow one of three crew cars that are available. They'll even give you a map to Tom's. The location makes this stop easily accessible from the Dallas, San Antonio, Austin and Houston areas, even in a Slowhawk like mine.

AMENDMENT

Freebirds Burritos, just across from the Texas A&M dorms, is the best burrito I've ever eaten. Well worth the short drive from the 'port into town. They are very easy to find. Just ask for direction at the FBO, where courtesy cars are usually available. Try to eat the "MONSTER," I dare you!

Corpus Christi (C.C. Intl. - CRP)

Astor's 

Although this is not a fly-in restaurant, it definitely qualifies as a place to check out. Astor's in Corpus Christi is a must for those who enjoy great steaks. We flew into CRP and some friends took us there for dinner. The steaks were tremendous, and inexpensive too! A large ribeye, cooked to perfection, was $7.95. Five adults consumed four steaks and some stuffed flounder for a little over $60.00. Astor's is not too far from the airfield. It only took us about a 5-10-minute drive, after we were done stuffing ourselves to get back to the airport. CRP is a large field served by major airlines. Taxis are readily available.

Corsicana (Corsicana Municipal Airport - CRS)

Fitzgerald's Trading Post 

Look out your left window as you turn final for runway 14 and you'll see it, Fitzgerald's Trading Post, just one mile north of field. Good BBQ sandwiches and hamburgers. They'll even cook one for breakfast. A courtesy car is available at the airport. The smell of hamburgers cooking just hangs in the air and gets into your clothes. You can continue to enjoy it the rest of the day.

AMENDMENT

Fitzgerald's Trading Post is still at it with breakfast, BBQ sandwiches, and some huge hamburgers. A 10-inch bun, 2 lbs. of meat and 1/2 lb. of bacon. Hours are from 4:00 AM till midnight. Sometimes he doesn't get there till 5:00 AM if its raining. You won't find me out flying in the rain at 5:00 AM.

Cuero (Cuero Muni. - T71)

Wal-Mart 

Land at Cuero Muni (T71), tie down and hop the fence. Right there it is, the BIG gray building, a Wal-Mart. You have probably seen one before. The good news is they have food. A small lunch counter makes burgers and hotdogs. The price is low. The quality is best judged by your hunger level. The hungrier you are, the better the food. Now don't forget that Ol' Sam Walton was a private pilot. He started with an Ercoupe (good choice) and finished with a Piper Malibu.

Dalhart (Dalhart Muni. - DHT)

Airport Cafe

Dalhart has a nice little diner right on the field serving hamburgers and the like. It turns out to be a great fuel and food stop between Austin and Denver.

AMENDMENT

Boy! Talk about a great little airport for a stop! These have to be the friendliest people I've met in a while. We had to spend some time there last weekend for weather. The diner was great and very reasonably priced. I didn't try it, but others mentioned that they have the best pie in Texas. I can vouch for the sandwiches and fries.

Several motels are nearby (3-5 miles), and some will pick you up from the airport. The FBO has a satellite radar/weather system.

Dallas (Love Field - DAL)

Uncle Julio's Mexican Food 

Uncle Julio's Mexican Food, southeast of Love Field on Lemmon Avenue, was wonderful. The grilled jalapenos were the best and their enchilada dinner was very good also. The salsa tasted quite unique, but I am only going to be able to guess how good their margaritas are, since I was flying that day. We flew a Cessna Skylane into Dallas Airmotive and were transported via courtesy car to the restaurant, which was about two miles away.

Dallas (Redbird - RBD)

Casa Blanca 

The Casa Blanca restaurant is located under the control tower in the main terminal building. Just taxi up to the door. The restaurant serves breakfast from 8:00 AM to 10:00 AM, lunch

from 11:00 AM to 2:00 PM, and dinner from 5:00 PM to 7:00 PM. Sandwich service is available all day. My wife says the breakfast is better than Denny's.

Decatur TX (Decatur Muni. - 8F7)

Mattie's 

We went to Decatur Muni. and by shuttle went to Mattie's on the square. It is three miles from the airport and well worth it. The square is the town center, with the courthouse in the middle and all the shops around it. I had chicken-fried steak. All meals are under $5.00, and kids meals are $1.75. Next time, I'm going to try the ostrich burger, $3.95. They also have evening buffets for $7.00 to $11.00.

Decatur (Grass Strip - Private)

Skydive Texas 

I like to fly to a grass strip (33.16.00W/97.26.50N) located just outside of Decatur, Texas. It is 3,800 feet and is usually pretty smooth.

This is the home of Skydive Texas. They have a small restaurant that serves daily specials on weekends only. Good food, friendly-people and you can watch the jumpers after you eat, or go for a jump yourself! Be very cautious when landing, bodies tend to fall out of the sky!

AMENDMENT

Went to Skydive Texas (Decatur private) and wasn't able to eat, as the restaurant closes early.

Dilley (Dilley Airpark - 24R)

Dairy Queen 

If you are between San Antonio and Laredo, land at Dilley (24R). Park at the south end of the field and there is a wooden gate into the parking lot of the Dairy Queen. I go every once in a while. They serve a good hamburger.

Fort Worth (Alliance - AFW)

Sonny Bryant's BBQ 

I stopped in at Alliance Airport (AFW) and was given a courtesy car to get to Sonny Bryant's BBQ, just off the approach of 34. Food was great, moderately priced, and the FBO was very nice. I tipped them with an order of fried-onion rings. This airport is between Northwest Regional and Mecham and is a new destination for fly-in diners. Enjoy!

Cactus Flower Cafe 

If you go to Alliance Airport without making it to the Cactus Flower Cafe, you haven't eaten the best food offered. Huge hamburgers are my favorite; however, several home-cooked dishes are served, from bodacious chicken-fried steak to grandma's meatloaf, not to mention

the most mouth-watering rolls. To enjoy your own king-size feast, simply arrive at AFW and taxi to the nice, new, and modern FBO. They will either slip you the keys to the loaner or give you a lift to the restaurant.

Fort Worth (Hicks Airfield - T67)

Rio Concho Aviation 

Come visit the newly remodeled FBO, Rio Concho Aviation, at Hicks Airfield (T67) for the best fresh-made deli sandwiches, burgers and pizza. Hours are Tuesday – Sunday 10:00 AM - 6:00 PM. (817-439-1041)

AMENDMENT

We went to Rio Concho at Hicks Field and had some excellent hamburgers. Since you park your plane next to the restaurant, this is one of those true fly-in places! It's small but staffed with real friendly folk. Hours update: Tue.-Sat. 11-6, Sun 11-3.

Fort Worth (Spinks - FWS)

Cracker Barrel 

While many think Fort Worth Spinks is living on borrowed time — political intrigue is threatening to remove it from the map, a new development might lure sky-diners there. Within walking distance of the terminal, it appears that a new Cracker Barrel is under construction! Yes, friends, strap on your Continental-powered bucket o'headaches and convert some cash into noise! Park your heap at FWS and amble to good, reliable country cookin'.

Gainesville (Gainesville Muni. - GLE)

The Windsock Cafe 

The Windsock Cafe has now officially opened at the Gainesville Municipal Airport in Gainesville, TX. They are open seven days a week. Their hours are Monday - Friday 10:00 AM to 2:00 PM, and Saturday and Sunday they are open from 10:00 AM to 6:00 PM. They have a great variety of food, ranging from meatball sandwiches to smoked turkey. They also have baked potatoes, frito pies, taquitos and much more. Their weekday specials include turkey or ham sandwich, chips and fresh-baked cookie for only $3.10 plus tax! (940-665-9825)

Galveston (Scholes - GLS)

Rain Forest Retreat 

There are many reasons to fly to Galveston. All of them are good!

The beach is a short walk from the airport. If you like the Gulf and I do, that's reason enough.

On the field is the soon-to-be world famous Lone Star Flight Museum. The Lone Star Flight Museum by itself is enough to make the trip worthwhile. Tie down on their ramp and walk inside. A third great reason, and on the airport property, is Moody Gardens — a real oasis! They have one of the largest IMAX theaters in the world, really do. It was here that the first IMAX feature film was shown. Throw in an indoor rain forest and an outdoor water amusement park and you have the makings of a great afternoon.
There are three restaurants on the Moody Gardens complex and a couple of snack bars. You won't starve and the food is pretty good.

Gordonville (Sherwood Shores)

Pelican's Landing Restaurant 

Listed as Sherwood Shores on the DFW sectional, this is truly a fine reason to "run for the border." The restaurant is a nine-minute stroll south of the aircraft parking area at Sherwood Shores. It is part of the huge Cedar Mills Marina. Opened in 1984, the eatery has been flooded several times and was last rebuilt in 1993. Fly-in visitors can nearly hit the restaurant with a thrown stone from parked aircraft; however, a small inlet lies between the parking area and this delightful little restaurant. The restaurant is open Wednesday through Sunday, starting at 11:00 AM weekdays and 8:00 AM weekends. An attached open-air dining room (zipped in plastic on chilly days) overlooks the attractive marina, bristling with tall masts. Inside and out, the restaurant is spotless. Diners are greeted by numerous smiling staff. A good view is to be had from the dining room, where non-smoking customers are well away from their addicted brethren. We sat down at 10:00 AM, in time for weekend breakfast. Our waitress greeted us and poured my coffee simultaneously. No bilge brew, either; the coffee was fresh, hot, and endlessly refilled. Service was exemplary. The day's chalkboard breakfast special was smoked pork chops, two eggs, and biscuits, all for $5.95. My wife, KayCee, ordered it and was pleased. I had a nicely done plate of breakfast tacos. KayCee offered me a bite of homemade biscuit, which prompted me to order my own, à la carte. I accused the waitress of having someone's sweet, old grandmother manacled to the stove. She laughed, but I spotted a strange, haunting glint in her eye. With orange juice, tea and coffee, our total bill came to $13.81. Lunch and dinner are moderately priced. Burgers are less than $4.00 and dinners begin at $9.00. Hangarmate Dan Meeks, a frequent visitor, claims the lunch and dinner is just as delicious as breakfast. After eating, we walked across the street to the pristine Ship's Store and Boutique. A friendly clerk gave us brochures revealing that the resort has cabins, RV hookups, and a sailing school. There are plans, she said, to build duplexes near the airstrip. Then, enroute back to our waiting C-170, I let myself into the enclosure that houses the marina's pot-bellied pigs and ewe. The larger of the pigs is the size of a large oil-drum and will prompt you to rub her belly.

Taking off downhill, toward the lake, I was careful to ascertain there was no inbound traffic, landing opposite-direction. A lazy turnout over the water afforded one last spectacular view of the entire resort.

Special thanks to Chicken Hawk owner, buddy and hangarmate, Dan Meeks of Fort Worth, TX, for first discovering Sherwood Shores. Ten of us bust our knuckles together in Hangar 29, KFWS, which you might remember from the movie about Pancho Barnes.

AMENDMENTS

Flew to an airport just north of Cedar Mills Marina on Lake Texoma (TEXAS) last evening. It is not shown on the latest DFW sectional but should be at next printing. It is a grass strip running east/west and about 3,000 feet long. Do not worry about the lone tree on the east end;
you will have enough room. Plenty of parking, and on Wednesday through Sunday you can eat at the Pelican Landing Restaurant. It opens on Saturday at 8:00 AM for breakfast. Rich Worstell owns the airport and marina. He is a real nice gentleman and is a member of our local EAA 323. He has some forward looking plans for the aircraft side of the business. It is already a very nice marina. I've been there when I had a boat. Sold it because it was competing with my new airplane.

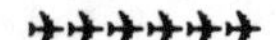

Pelican's Landing Restaurant at Cedar Mills Marina in Gordonville, TX, has a beautiful turf strip, marina, restaurant and club on the lake, with the best mushroom cheeseburger you will ever have anywhere.

Pelican's Landing at the Cedar Mills Marina on Lake Texoma is a fair, if ordinary grill (hamburgers, steaks, etc.), but the view is spectacular for those who like sailboats. (903-623-4077)

The public-use grass strip is not on recent sectionals (33 50.11N, 96 49.35W or 26.6 DME on the 133 radial from ARD VOR, SW/NE). The runway is 3,300 feet of smooth sod. You must walk about 1/4 mile from the strip to the restaurant.

Cedar Mills is just like everyone says it is! Rich Worstell, the owner, flies a Shrike Commander and a Citabria with his son. He keeps the runway in great shape. We landed three Bonanzas in a row with plenty of room. There are sailboats to rent, and there are a lot of sailors who will take you for a spin just by walking to the dock. The phone number has been changed to 903-523-4200.

Graford (Possum Kingdom Airport - F35)

Lefty's

Approximately 70 nm west of Ft. Worth, there is a quaint little BBQ place on Possum Kingdom Lake. They have fantastic BBQ, fried fish and BBQ sandwiches. I have eaten here two times and have said I would go back each time.

The BBQ place is about a mile and half from the airport. There are flyers and a menu in the pilots' rest area. The first time we flew in, the owner's wife came and picked us up just after finishing all her baking at home. We had to ride the mile and half with eight home-baked pies! I tried the pecan pie after lunch; it was great. They wrapped up two BBQ chickens that we carried home with us.

The second trip, the owner came and picked us up at the airport. The owner is the fire chief and works with the county judge to help folks serve out their "community service" time. He had a couple of young men ticketed for excessive speeding washing and waxing fire trucks the day we were there. The food is great, reasonably priced, and the owner and his wife will tell you a story or two.

Graham (Graham Airport - E15)

Dairy Queen

There is a nice clean Dairy Queen at the Graham Airport (E15). The DQ is right across the street. The FBO does have a courtesy car should you need to go into town. Fuel is reasonable and service is friendly. My wife and I fly up there often from our home base at the Arlington Municipal (F54). Give it a try. Nice airport!

Granbury (Granbury Muni. - F55)

The Nutt House

Granbury (F55) is one of those "instant antique" towns, a retail destination filled with tourists and cluttered craft shops hawking everything from grandfather clocks to German beer steins, authentic and otherwise. Like many small Texas towns, it has a central square and an ornate stone court-house; this one figures prominently in several Texas-related wars. In addition to advertising its wares around the DFW metroplex, Granbury has accelerated its prominence by featuring its accessibility via car, train (a scenic ride from Fort Worth), and airplane (a nice little airport with a VERY friendly FBO).

The blurb for this restaurant, The Nutt House, indicated that someone from the restaurant would pick us up at the FBO. It was the sultry Saturday afternoon of Labor Day when we went, so the place was packed. On arrival at Granbury Muni, we called the restaurant, and the woman we spoke to said they were swamped and asked us apologetically to find our own ride, if possible. If we could not, she'd be out in about half an hour.

So we sauntered to the counter and were about halfway through asking about taxicabs when the man running the FBO handed us a set of keys to an official Granbury vehicle! The white sedan had the plates you see on cop cars, a big sheriff-style star painted on the doors, "City of Granbury," and the big searchlight on the driver's side. My friend grabbed the keys, said, "You flew. I'm driving!" and jumped in. We figured that with this rig, we could park anywhere but we didn't push it.

The Nutt House restaurant — it's an old family name, Jess and Jacob Nutt — has plenty of character. It was built in 1893 and is decorated with period furnishings. It's listed in the National Register of Historic Places. When you walk in, a wall-length 19th-century scroll actually traces man's genealogy from Adam to Jesus to modern 1878, with many cultural notes on the time scale. Other artifacts are equally engaging.

The cooking is country-style, natch. It's always a buffet, with all you can eat. This includes salad bar, several meats, three vegetables, various fresh breads, and two desserts. As a cultural experience, you can't beat it. The Nutt House's phone is 817-573-9362.

Grand Prairie (Grand Prairie Airport - GPM)

Final Approach

The Final Approach is located in the terminal building at Grand Prairie Airport (GPM). There are lunch specials during the week. Hours are 7:30 - 3:00 Mon. - Wed, 7:00 - 9:00 Thur. – Sat., and 8:00 - 3:00 Sun. The food ranges from burgers to steak, with pie and cobbler available for those who aren't too picky about their weight.
AMENDMENTS

Flew down to Grand Prairie on Monday, December 11, 1996. The restaurant is closed. The controller in the tower told me that the city was trying to get someone else to take over the restaurant.

The Final Approach has reopened at Grand Prairie Airport. The menu is pretty much the same, but changes by the new operator are in the works. New furnishings are very comfortable. I have eaten there five or six times now and the food is very good. There are lunch specials on the weekend. At the present time, they are only open for breakfast and lunch, but there are plans for a dinner menu.

Greenville (Majors - GVT)

Mary's of Puddin' Hill 

Last Saturday we flew into GVT for lunch. I used their very nice courtesy car, a late-model Caprice with a "certified speedometer." Mary of Puddin' Hill is about seven miles from the airport. The deli was very good. Their desserts were heavenly!

Hilltop Lakes (Hilltop Lakes - T38)

The Club House and Restaurant 

Sunday brunch at Hilltop Lakes Resort is delightful. The runway, 16/34, is compacted iron ore and smooth. The Club House and Restaurant is a short walk. Just across the street from the airport, it sits on a hill overlooking a lake. Big bay windows provide great views of the countryside from the dining room. Get there before noon, because the church crowd can get large. Fly over the airport first and talk on 122.75 before landing. It is listed as private, but they love to have visitors. Don't be surprised if you see several aircraft parked on the ramp. Telephone is 409-855-2222.

Hondo (Hondo Muni. - HDO)

The FlightLine Cafe 

Landing at Hondo Municipal Airport is like taking a trip back to 1942. The airport is dotted with wooden barracks and Quonset-style hangars. There are five extra-long 150-foot-wide paved runways. In the center of 17R/35L stands a building which was once part of the navigator training complex. Today, it is The FlightLine Cafe, owned and operated by Betsy Hermann. It doesn't look much different from the officer's mess of 50 years ago.

Inside it is stacked with memorabilia. All the cadet class books, beginning with August, 1942, are here. There is a photo collection that will really take you back.

The FlightLine Cafe has a wide selection and is particularly famous for is baked goods: pies, cakes, pastries and breads. I was immediately attracted to the TEXAS-SIZE chicken-fried steak. There is also a terrific buffet for a mere $4.95!

AMENDMENT

After checking the $100 hamburger web page, my wife and I embarked for The FlightLine Cafe. Hondo came through—from the salad bar made out of a WWII jeep to the old wooden prop hanging just inside the door. We looked over the old yearbooks and read about the 55-year-old training field's history. The people were friendly and the food was Texan, with aviation names like "the Audie Murphy burger" and the "Army-Navy" plate. We met another pilot who flew in a homebuilt VariEze to meet some friends for lunch.

Hours of operation: Monday - Thursday 8:00 AM to 2:30 PM, Fri. 8:00 AM to 9:00 PM, Sat 8:00

AM to 2:30 PM. Restaurant phone number: 210-426-4020.

Houston (Andrau - AAP)

Cliff's

This is one of Houston's really GOOD hamburger grills. What's nice about this one is that it is a short half-mile walk from Houston's close-in west side Andrau Airport. You land, tie down at the terminal and walk straight down the only road.

You'll cross Richmond Ave., a four lane. Then at Westheimer, another four lane, you cross and jog a hundred yards to the left. There it is, Cliff's. Enjoy! I had the charbroiled chicken breast with a made-from-real-ice-cream vanilla malt.

The unhappy part of the trip comes when you get back to the airport office. They will want to collect a $5.00 landing fee unless you buy fuel.

Houston (David Wayne Hooks - DWH)

The Aviator's Grill

The Aviator's Grill is a very friendly restaurant located directly on the airport grounds adjacent to the fuel services (north end of the airport). It has a great view of the runways. It's very clean and is decorated completely with "golden age of aviation" murals and other decorations, which make for a very pleasant "airplane watching" environment. The people that run The Aviator's Grill are extremely friendly and the service is good. The food is inexpensive and well prepared. There isn't a huge menu, but most of the basics are there, ranging from the half-pound "B-52" burger with fries at around $5.00 to chicken-fried steak with mashed potatoes and white gravy at $6.95. They have several different sandwiches as well as fried catfish and fried chicken. There is also a daily special. They are open for lunch during the week (11:00 AM to 2:00 PM Monday through Thursday) and lunch and dinner Friday and Saturday (11:00 AM to 9:00 PM Friday and Saturday). In summary, it's a very friendly place with a good view that will serve you a good meal for not a lot of money. Recommended. Remember, David Wayne Hooks Airport has a sea lane, so you "float flyers" can feel welcome, too!

Houston (Weiser Air Park - T17)

Carl's Bar-B-Que

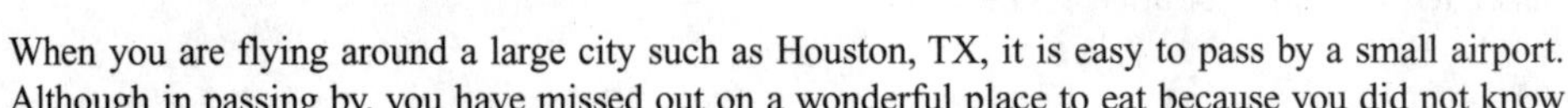

When you are flying around a large city such as Houston, TX, it is easy to pass by a small airport. Although in passing by, you have missed out on a wonderful place to eat because you did not know it was there. Such is the case with Carl's Bar-B-Que.
Located along US 290, northwest of downtown Houston and Beltway 8, is Weiser Airpark. If you are flying in the Houston Class B airspace (with a clearance, of course!), Weiser Airpark looks small from altitude. It is only after a closer look at the charts that you realize you can land there, no problem.

I had an occasion to visit Weiser the other day. My friend invited me along for a ride in his Ercoupe. Now this was a delight in itself. About lunchtime, he suggested we stop at Weiser. I remarked, "Well that's fine, but where do we eat?" It was then that I was educated that a great place called

Carl's Bar-B-Que resided right next to the airport property.

After landing and walking around the airport for a short tour, we walked a very short distance to Carl's. Well, it was worth every step. This restaurant had barbecue from sandwiches to plates lunches, sure to please any Texas-size appetite. If you are out flying, consider stopping at Weiser for a bite to eat at Carl's. You will not be disappointed!

AMENDMENT

Being a local, I have several choices of eateries, what with Houston boasting more than 2,000 restaurants. Carl's Bar-B-Que, butting up against Weiser Airport, no kidding...about a 120- second walk from tie-down, ranks up there with classic Texas barbecue's. My family enjoys the stuffed baked potato above all else. It comes in three varieties, the largest being heaped, I mean overloaded with fresh chopped barbecue. I have a big appetite, yet am frequently unable to finish it. Enjoy!

Huntsville (Huntsville Muni. - T39)

New Mount Zion Missionary Baptist Church 

This is the best BBQ feed in the world. Period!

We have never before awarded six burgers, but this place rates that many and more. Just 75 miles north of Houston, it is worth a detour and a landing on any business day and a special trip on Saturdays. This is the "church of the holy barbecue." It has been smoked by an angel named Annie May Ward. She is the head pit diva at this decidedly down-home place.

What started 16 years ago as a one-time way to raise money for the congregation of this east Texas church has become a for-real Q shrine. It is open from Tuesday through Saturday. For seven bucks you're served ALL you can eat.

Call Huntsville Aviation before you takeoff and reserve the airport car, 713-295-8136, or ask them to call a cab for you when you're 10 miles out.

Kennedy (Karnes County Airport - 2R9)

Barth's Restaurant 

The FBO can be counted on to fill or fix your airplane and/or give you a ride to Barth's. It's more fun just to park and walk through the weeds the quarter mile to this friendly place. Barth's is a good old highway cafe, famous for its chicken-fried steak.

Killeen (Killeen Muni. - ILE)

The Airport Coffee Shop 

Killeen Muni. (ILE) is a very nice airport if you don't stray into the Ft. Hood airspace just north of the field. The terminal building is the site of the Airport Coffee Shop. The name says it all. It is a coffee shop at the airport. They serve pretty standard fare. It is well worth a stop when you and the ship need fuel.

Lampasas (Lampasas Muni. - T28)

Storm's

Want a REAL hamburger? Try a "Storm Burger" at Storm's! It's more than you can usually handle, but try! I can't say there's a shuttle running, but if you ask Bob Brame, the FBO, he'll see if he can get a ride for you. It is only about 1.25 miles south of Lampasas County Airport, so you could walk it!

Lancaster (Lancaster - LNC)

Happy Landing Cafe

OK, here's the scoop. At Lancaster, Texas, they've got good fries and burgers — cheap, fast and the portions are generous. The place is right on the field, plenty of airplane parking. I guess at least four burgers. Go check it out!

AMENDMENTS

Well, the text for the Happy Landing didn't really do it justice. I live in Plano, TX, and have visited the Happy Landing Cafe almost every weekend since my friend (the pilot) arrived. It not only has GREAT! hamburgers and fries, it sports a menu that has breakfast and a full lunch offering. The hours are Tuesday - Saturday 7:00 AM - 3:00 PM, Sunday 7:30 AM - 3:00 PM, closed Mondays.

The other neat part is that you can taxi almost to the door!

Lancaster (LNC) is a great place to stop for lunch, friendly people, good food. The club sandwich is excellent, and an excellent runway.

Livingston (Livingston Muni. - 00R)

The B&B Bar-B-Que

I fly to Livingston, TX, for the best barbecue that I have ever had. The B&B Bar-B-Que is owned by Leon Barnes. It is within walking distance of the airport. From the outside it may not look like much, but the food is great. The last sets of hours that I had were Mon.-Tue. Closed, Wed.-Thur. 1030–1900, Fri.-Sat. 1030 – 2100, Sun. 1030-1800

If you want to call ahead of time to check on the hours, the phone number is 409-967-4108.

Livingston Airport has one runway. Be cautious, runway 30 does not have a taxiway, so planes may be back taxiing on the runway.

Llano (Llano Muni. Airport - 6R9)

Cooper's Bar-B-Que

Take your best appetite along when you land at Llano, TX Municipal airport. Call ahead on Unicom, ask Evelyn for the courtesy car and take the short drive into town to Cooper's Bar-B-Que. They employ an unusual, but delightful serving routine. Largest outdoor set-up I've seen serving ribs, beef, sausage, pork chops, chicken and sirloin steaks from the B-B-Q.....check your weight & balance before take-off! Our gang rates it 5 ribs! No burgers spotted around this place!

Lockhart (50R)

Chisolm Trails Bar-B-Q

For good Texas barbecue and extra-friendly people, fly to Lockhart. Debra, the owner of Drasur Aviation, has a Suburban Airporter courtesy car that will carry SOME folks. So bring your friends and go about a mile toward town from the airport to Chisolm Trails Bar-B-Q for some of the finest food around. Phone 512-376-9608.

Lufkin (Angelina County - LFK)

Airport Restaurant

My home airport at Lufkin, TX, is Angelina County (LFK), located in the deep east Texas piney woods. This newly refurbished airport terminal is the setting for a terminal cafe which is attracting a lot of attention. Everything from home-cooking to the best "terminal burgers" made with THICK beef patties are served up by Mildred. Line service is excellent. A real show-stopper is Mildred's famous "governor's choice" candies. Imagine a peanutbutter cup half the size of your fist! So good, Gov. Richards is rumored to have flown all the way in from Austin just to replenish her supply once. All the conveniences of a big-city airport (VOR, multiple instrument approaches, E-W & N-S runways, professional service, superior food), yet sandwiched in the midst of national forest land and close to Sam Rayburn Lake. Excellent x-country escape.

McAllen,Texas (McAllen Muni. - MFE)

Tony Roma's

I flew down to McAllen Miller (MFE) today. Parked at McCreery Aviation, which has been in business 50 years. They have a very interesting picture wall. Everybody from Bob Hope to George W. Bush. Tony Roma's Ribs is just across the street from McCreery. There is a mall just down the street, also, if you want to shop.

McGregor (McGregor Muni. - F60)

Convenience Store

McGregor Muni, four miles west of McGregor, TX, is a very nice little airport with the friendliest people you'd ever want to meet. The FBO is clean and comfortable, with all the amenities: pilots' lounge with TV, VCR, movies, and a place to sleep if you need it; a flight planning room; pilot supplies; maintenance facility; and all the free popcorn, ice tea, and coffee you care for. Just off the entrance to the field is a small convenience store with burgers and such, all very good. The FBO is manned from about 7:00 AM to around 9:00 PM. Fueling is full service, and I mean the old-fashioned kind of full-service-with-a-smile service. This place is well worth the visit.

Mexia (Mexia-Limestone Cnty. - TX06)

The Tamale Inn

A friend of mine and I flew down to Mexia to try out the lunch fare at the Tamale Inn. I drove by the restaurant several months ago and noticed an airport right across the street. The Mexia Airport is located about 50 minutes south of the DFW area and is ~25 miles due east of Waco. After we landed, we parked the plane at the FBO, which is located at the north end of the runway. The restaurant is located at the south end of the runway and across the street. It took us about 10 minutes to walk there from the plane. They did have hamburgers, but I decided to try out the $100 fajita plate special. My friend had a $100 taco salad. They gave us the chips and the hot sauce along with plenty of iced tea. They also have beer on the menu for the passengers. The fajitas were well seasoned and came out sizzling hot. Both the fajitas and the taco salad were rated excellent to outstanding.

I told another friend of mine about the place, and the next day he flew down with his wife and family. They got out of the plane at the FBO. The FBO saw my friend with his four children and let them use the courtesy van to get to the restaurant. The operator told my friend that he was closing for the evening and to leave the keys to the van under the seat when he returned. They also rated the food as excellent.

The FBO also told my pilot friend that sometimes hungry pilots will park their planes at the south end of the runway and hop the fence to save them the walk to the restaurant.

AMENDMENT

I flew there on Sunday and was frustrated to find they had closed 10 minutes before I arrived. So that no one else will have the same experience, their hours are: Open 11:00 AM - 2:30 PM and 5:00 PM - 9:00 PM Mon. through Sat. and 11:00 AM - 2:30 PM on Sunday.

Mineral Wells (Mineral Wells Muni. - MWL)

Woodie's

There is a hamburger restaurant in Mineral Wells, TX, that has the BEST hamburger in the state. It is called Woodie's and is located about three miles from Mineral Wells Municipal Airport (MWL). There is a courtesy car available. If it is already in use, you can catch a ride with a local or walk.

Mt. Pleasant (MSA)

Bodacious BBQ

Mt. Pleasant, Texas, is a very "pleasant" town east/northeast of Dallas. It is about 45 minutes to an hour in a 172. The airport has a nice, 3,800-foot strip and an FBO that reminds me of the Petticoat Junction television series. It has 100LL self-serve.

The people at Bodacious would not, or perhaps could not come pick us up at the airport. We called Everyday Cab at 903-572-3623. For $5, they were prompt and very friendly. Bodacious BBQ is first-class BBQ. I had a beef/sausage combo plate that was mouthwatering. The homemade cornbread was out of this world. Don't bother with the fried pies for dessert, however. They were more like glazed donuts filled with jam rather than traditional fried pies.

Nachogdoches (A. L. Mangham Jr. Regional - OCH)

Clear Springs Restaurant 

(Och) AWOS 409-564-5074. The Clear Springs Restaurant has decor of years ago. It was a real ice house next to the train station in the oldest city in Texas! The fried catfish is great, and the onion rings, their specialty, are made from scratch. Red Air is the FBO and can generally get you to town. If not, a taxi is $6.00 and worth it. For us IFR guys, it has a full-blown ILS.

New Braunfels (New Braunfels Muni. - 3R5)

The Cloud Nine Restaurant 

New Braunfels is the southern gateway to all the pleasures of the Texas hill country. The Cloud Nine Restaurant is in the terminal of the more-than-adequate New Braunfels Municipal Airport. The doors are open from 7:00 AM - 3:00 PM Monday - Friday. The schedule changes a bit for Saturday and Sunday, 7:30 AM - 4:30 PM. This is a very worthwhile stop for breakfast, lunch or just a cup of a coffee. Give the manager, Pamela J. Roe, a call at 210-629-1700 to arrange aircraft catering. We like it so much that we sometimes fly all the way from Houston just for breakfast!

Palacios (Palacios Muni. - PSX)

Palacios Mexican Restaurant 

The pilot's lounge is open Mon.-Fri. 7:00 AM to 4:00 PM. A pay phone is available to call local restaurants. Take a quarter for the pay phone and a hearty appetite to Palacios Mexican Restaurant, 511 Main. Daniel Molina serves up homecooked meals that won't leave you hungry. I recommend the shrimp fajitas. Open seven days a week 11:00 AM to 9:00 PM, Friday and Saturday nights till 10:00 PM. Call for airport pickup, 512-972-2766.

Petersen's Seafood Restaurant 

I've flown into PSX about a dozen times over the past couple of years, mostly to eat at Petersen's Seafood Restaurant. Excellent fresh seafood! The pilot lounge has a pay phone on the outside wall. On the wall next to the pay phone is Petersen's phone number. Mr. Petersen will come down and give you a ride to the restaurant and back when you're done.

Pampa (Perry Lefors Field - PPA)

Dyer's BBQ 

Make use of the courtesy car at the FBO and drive south to Dyer's BBQ. Good food and good company!

Coney Island 

Here's an amendment from someone who has eaten a lot at the Coney Island in downtown Pampa. The same family fixes great coneys, wonderful stew, chili and fabulous homemade pies. We used to carry them back to the Metroplex by the bag full and freeze. There's a doctor in Amarillo that does the same and keeps them in his office freezer. Definitely worth a trip to Pampa!

Pecos (Pecos Muni. - PEQ)

Swiss Clock Inn

Another good place to go eat is in Pecos, TX. Fly in to the Pecos Municipal Airport, where the service is friendly and they have an airport car you can borrow. Just a short distance away is the Swiss Clock Inn, which has a pretty decent restaurant. A good place to stop in the middle of west Texas!

Port Aransas (Mustang Beach Airport - 2R8)

Pelican's Landing

Port Aransas is located on Mustang Island, a barrier Island about 18 miles offshore from Corpus Christi, TX. This is the place to go for great fresh seafood ! The city provides a FREE motorized trolley that runs from the airport to downtown, to the beach, to the birding center, to the park and back to the airport every 50 minutes. The trolley runs seven days a week from 10:00 AM until 5:30 PM. Ring the bell to get off. Flag the driver anywhere to get on. Pelican's Landing is a super place, offering a Cancun atmosphere. Try the shrimp.

Roanoke (Northwest Regional - 52F)

Pappy's

Pappy's is at Northwest Regional (52F). Supersized burgers, greasy fries and coffee or iced tea for $6.00. An easy walk from anywhere on the airport.

AMENDMENT

Pappy's hours are 0830 to 1600, Tues. - Sun., closed Mondays. It's located 1/4 mile west of the airport at approximately midfield. Owner, Danny Rascoe. Phone, 817-491-1009. I've eaten at this place on a few occasions and am always amazed at the quality of the food for the price. I almost feel guilty paying so little.

Robstown (Nueces County - T53)

Joe Cotten's

Fly 15 nautical miles from Corpus Christi, TX, on the 231 radial and you'll find the Nueces County Airport. It has a well-maintained, lighted, 3,050-foot hard-surfaced runway. There are a few hangars, an FBO and fuel.

I came to eat not visit! The best barbecue joint in the civilized world, Joe Cotten's, is here. Find the phone, drop a quarter and dial 512-767-9973. Within minutes, the FREE transportation wagon will arrive.

Cotten's building is as rustic as his waiters are eager. You'll be shown to a table immediately! There's a large selection of meats that have been slow-cooked over mesquite wood. If you expected to be served on a plate, you're out of luck. Here, they do it differently. White butcher paper will be spread before you. Meat and condiments will be slapped on it.

Q sauce is presented but not needed. The meat is choice and well seasoned. I had a chopped beef sandwich, beans and salad, the best I have ever eaten for a reasonable $5.95, iced tea included. Full meals go for $7.95. Be warned, they are HUGE!

Joe Cotten's has been doing it right in south Texas for more than 40 years. Experience accounts for the quality of his product. Cecil Cotten is the host these days. Pilots have been flying here since his daddy started all those years ago. We wish Cecil a lot of luck keeping the legend alive.

AMENDMENT

Went to Joe Cotton's in Robstown, Texas, from Laredo, Texa,s and it was everything the $100 Hamburger said it would be.

Rockport (Aransas County - RKP)

The Sand Dollar Restaurant 

The Sand Dollar Restaurant in Rockport offers excellent seafood in a beautiful setting. The restaurant is built out over the bay in a marina setting. Food prices are reasonable (far lower than the Houston area), and the service is good. The menu has a very good variety of both seafood and landlubber entrees. The restaurant is a couple of miles from Aransas County Airport (RKP), and transportation to/from the Sand Dollar is complimentary. RKP has NDB and VOR-DME instrument approaches and makes a really nice stop on cross-country flights.

As you approach RKP, ask the Unicom operator to call the Sand Dollar for you. The taxi will be waiting when you finish your shutdown.

Who/Dat 

The Rockport taxi, 512-729-8294, charges $3.00 to take you to the Who/Dat restaurant. It is named after current Aggie football player Dat Nguyen and operated by his family. They serve lots of good fresh seafood.

San Angelo (Mathis Field - SJT)

Airport Cafe 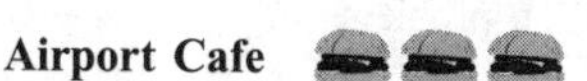

Mathis Field has commercial air service and a real airport restaurant in the terminal building. Enter through the door on the left. I don't know the hours, but there is enough traffic so that it's probably open throughout the week.

AMENDMENT

The airport restaurant has primarily Chinese food, all of which is pretty good. They also have the requisite chicken-fried steak. I love the irony of a Chinese restaurant at a smallish west Texas airport.

San Antonio (San Antonio Intl. - SAT)

Bill Tasso's Barn Door 

This site is not on the airport, but it's close enough and the filet migon is unquestionably worth the effort. I have traveled the world in 30 years of flying and have not found another filet mignon quite up to the size and quality AT ANY PRICE as you will find at Bill Tasso's Barn Door. I've made this recommendation to many corporate drivers and so far, they have all come back to say thanks for the recommendation.

Exit San Antonio Intl. via the main entrance (going south), cross over I-410 and turn left (go east) onto the feeder, then onto I-410 and continue east for approximately one mile. Exit to the right at Broadway. At the stop light for the intersection of the feeder road and Broadway, you will see a street that makes an approximate 30 degreeangle to the right (southeast). Take that street for about 1/3 of a mile until you come to Nacadoches Road. The Barn Door is on that corner, on the right-hand side.

You simply can't beat the Blue Ribbon Filet but if you are REALLY hungry, they have a 72-ounce steak which, if you can eat it in an hour, is FREE!

Oh yeah, dress code can be anything from formal to boots and jeans (typical San Antonio). The prices are most reasonable.

San Antonio (Stinson Municipal - SSF)

Stinson Field Coffee Shop 

Stinson Field in San Antonio has a restaurant in the terminal building. Unfortunately, it's only open during the week, not on the weekends. They serve standard airport fare. Stinson is a nice tower-controlled field, and the old stone terminal building, with the control tower upstairs, is a pleasant reminder of a simpler era in air travel. You almost expect to find a DC-3 waiting out on the ramp.

Sherman-Denison (Grayson County - F39)

Grayson County Restaurant 

Hours: Monday - Sunday 8:00 AM - 6:00 PM. On the airport, you can park your plane within 25 feet of the front door. Breakfast is served anytime! The usual burgers and sandwiches are offered at excellent prices. Good place, reasonable with no ground transportation problems.

AMENDMENT

The Grayson County Restaurant at the airport is under new ownership. David Ledden is the owner, cook, and commentator. Hours are from 7:30 AM to 6:30 PM seven days a week, though he is frequently open till 9:00 or 10:00 PM. It's located at the south end of the field across from the Grayson flying FBO. A great place to meet local pilots, homebuilders, etc. Recommended highly.

Sonora (Sonora Muni. - E29)

Devil's River Inn

When I purchased my airplane, a 1947 Cessna 140, in Bakersfield, CA, I had to fly it back to my home base at the time in Houston, TX. On my second night, I landed at Sonora, TX (E29). A friend of mine had recommended I check out the little steakhouse located practically on the field: Devil's River Inn (Sutton County Steakhouse). Sure enough, the steaks are terrific! What's more, I doubt I paid more than 10 bucks for it. I made a point of keeping their business card. (915-387-3516)

Sulphur Springs (Sulphur Springs Muni - SLR)

Red Barn 

The Red Barn is a short walk off airport. It offers good food at good prices and no transportation problems.

Stephenville (Clark Field Muni. - SEP)

Stephenville Mexican Restaurant 

You might want to check out Stephenville, TX, Clark Field. There are three restaurants at the entrance to the airport, about ½-mile walk from the ramp area. One is a hamburger joint that is pretty good. There is also a Mexican restaurant and one other place.

Taylor (Taylor Muni. - T74)

Louie Mueller's BBQ 

Louie Mueller's BBQ in Taylor, TX, is one of the three best in the state. Awesome food! Get there early, definitely before noon on Saturday, or the food may be sold out and the place will close. If you miss out, just walk a block to Mikeska's and pig out. It is excellent as well, but not prone to sellout like Mueller's.
I borrowed a truck from a very nice fellow at Centex Flying Service for the two-mile drive into Taylor. Picked up a couple quarts of Aeroshell 100 when I got back. Nice folks and bulletproof BBQ.

Tyler (Tyler Pounds Fld. - TYR)

The Airline Restaurant 

Burger King has the Whopper, and McDonald's the Big Mac. Every fast food chain claims the best hamburger. Well, fellow aviators, there is an ULTIMATE. Fast food chains, eat your hearts out, for the hamburger I had the other day was truly the biggest I have ever seen.

My sandwich quest took me to the northwest edge of Texas' "Big Thicket." I landed at Tyler's Pounds Field. The Airline Restaurant is on the west side of the terminal building. Carol Dennison has owned and operated this monument to ground round since 1982.

Any world-class contender for the best burger title must have a name. Carol calls her's the Pounder, in honor of the airport. This is a Texas-size sandwich, a whole pound of ground cow, served on a 9-inch homemade bun. The Pounder comes on a megaplate and is cut in quarters. You can order a half plate or even a quarter, but big appetites demand the whole thing.

Your order comes with french fried potatoes or onion rings. I ordered half and half. A second plate came piled high with them. A third platter comes with all the trimmings: tomatoes, onions, lettuce and lots more. I am darn glad I brought a friend along. No way could I have eaten all that food, not in one sitting anyway.

The Airline Restaurant is open seven days a week. They are open Mondays through Thursdays 6:00 AM to 8:00 PM, Fridays 6:00 AM to 9:00 PM, Saturdays 6:00 AM to 8:00 PM, and Sundays 8:00 AM to 5:00 PM. Special groups are accommodated after 5:00 PM on Sundays call in advance for reservations. The telephone number is 903-597-3182.

I was really impressed by the hospitality of the people. When we arrived, the place was packed. Above the conversation and noise, a group already seated noticed us and my two 200 pounds of cameras. They were spread out comfortably over three tables. They proceeded to squeeze up onto two tables! They gave us the empty one, as if this was normal and expected. I was amazed! Coming from the BIG city, it was something that I wasn't used to. I thanked those folks at least three times. That's what I love about the South.

The best time to visit Tyler is October as we did. That's when the Rose Festival takes place. Tyler is known nationally as the City of Roses. It supplies 1/3 of the country's florists with pretty petals. About 100,000 people turn up for the annual event. This was the best year since it started in 1933. The Rose Garden Complex had just opened, housing its new $1.5 million Rose Museum. The complex includes a convention center, meeting rooms, and a gift shop. So go for the Roses, but don't leave without sampling the Pounder!

AMENDMENT

OK, you haven't lived until you have had the Tyler, TX, Pounds Field Pounder! Only a short flight from Dallas, a must-visit spot.

Victoria (Victoria Regional - VCT)

Leo's Feed Lot

Leo's Feed Lot in Victoria, TX, is a fantastic place to eat. Victoria Regional (VCT) is an old Army Air Corps base and spread over a few acres. Leo's is located on the west side of the airport property about ¼-mile walk from Victoria Aviation Services. The food is awesome! They serve steaks, seafood, burgers. You name it, they got it. I've not been disappointed by their cooking yet.

AMENDMENT

My wife and I just flew to Victoria (VCT) looking forward to reviewing the food at Leo's but found to our dismay that they are closed on Sundays. As in any flight, it pays to plan ahead.

Daniel's

The FBO is managed by the county now. Daniel's, at the terminal building, offers one of the very best hamburgers in town as well as delightful other dishes every weekday, by a man who loves to cook. Leo's, across the field, is still the grand cafe of the airport.

Waco (Waco Regional - ACT)

Airport Snack Bar

I ate here while waiting out thunderstorms. It's more of a snack bar in the terminal than anything else — burgers and sandwiches. The BLT was pretty good, though.

There's also an NWS station and a control tower that is open to visits to pass the time if you're weathered in like we were.

The Red Barn Steakhouse

The Red Barn Steakhouse is only about two miles from the airport and serves a very good steak.

The Hickory Stick

If you stop over in Waco, you need to eat at The Hickory Stick. This is a decades-old "institution" just about three miles up 19th Street from ACT. Very reasonable prices and great food. Borrow a car from Texas Aero and make the short trip.

Waller (Skydive Houston Airport - 3XS7)

Airport Sandwich Shop

Skydive Houston Airport is a neat place on a Saturday or Sunday to eat at their sandwich shop and watch the skydivers jump 24 at a time. There are two landing strips, one blacktop and one grass (the grass is better than 3,000 feet). It is located 30 DME west of Houston's IAH outside of the Class B airspace. If you like golf, a nine-hole course is next to the runway, and membership is not required.

Weatherford (Horseshoe Bend - F78)

Horseshoe Bend Restaurant

Horseshoe Bend Country Club Estates is mostly a weekend retreat and retirement community of mobile homes located 12 NM SW of Weatherford on the Brazos river. The FAA designator is F78.

I own a lot with a mobile home near the north end of the airstrip. The strip is mostly gravel with some grass. The club is about a quarter of a mile west of the airstrip. The airport is unattended; however, usually there is someone around the north end of the strip that will give visitors a ride to the club. The club also has a 9-hole golf course that is right off the north end of the airstrip.

There is a large pecan tree on final approach to runway 17 as well as a TV antenna. On the south end is a rather large hill, but the runway is approximately 3,500 feet long, so no problem. The club house restaurant opens at 7:00 AM and closes around dark. The menu includes the normal things that golfers like, breakfast, burgers, fries, and club sandwiches.

Horseshoe Bend is sometimes confused with Horseshoe Bay which is located near Marble Falls, TX, in the hill country. The name is the only thing they have in common.

AMENDMENTS

I visited Horseshoe Bend, and the restaurant has temporarily closed.

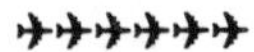

The restaurant at Horseshoe Bend Estates (F-78) near Weatherford, TX, is open again.

Weatherford (Parker County - WEA)

Driver's Diner

One of my favorite fly-in restaurants is Driver's Diner which is part of Driver's Truck Stop near Weatherford, TX. Driver's Diner is open 24 hours a day, 7 days a week, and is located adjacent to the Parker County Airport (WEA) which is paved and lighted for around-the-clock operation. Driver's Diner offers the usual truck stop atmosphere and fare. Sunday mornings can get interesting with the number of different aircraft that show up for breakfast.

AMENDMENT

We went to Driver's Diner on Saturday. The lunch buffet was all-you-can-eat chicken-fried steak, chicken, and barbecue (all homemade) for $5.95. This is not your ordinary truck stop. I was pleased and stuffed. They had some really great looking pie. Too bad I didn't have any more room to put it. The people were friendly, too.

The runway is narrow and not very level, so those that fly out of the nicer airports will get to practice some landing skills.

Weslaco (Mid Valley Airport - T65)

Arturo's

Progresso, TX, and Old Mexico can be reached by flying into Weslaco, TX, Mid Valley Airport (T65). Here's how it works. Mid Valley Airport is city-owned, with a nice terminal and friendly people that will call you a cab to go to Progresso, TX. It's about $10.00 a head and a 10-minute ride. Then you walk across a very short bridge to the busiest border town in Mexico. There are two nice restaurants for Americans. One is on the left called Arturo's, and the next one is one block farther down the main drag on the right called Gracias. It has a mariachi band for listening pleasure. The food is excellent and the restaurant is beautiful. This fly-to eating spot is a 10-star trip.

P.S. You can buy all of your American-made medicine for pennies on the dollar. Ask the border guard on the U.S. side what you can bring back across.

Wichita Falls (Kickapoo Downtown Airport - T47)

Gene's Hamburgers

Gene's Hamburgers has been in business for many years. Just fly into Kickapoo Downtown Airport (T-47) and use the courtesy car that is available most of the time. The FBO will give directions to the restaurant. If you fly in around noon, be prepared for a short wait. The fries are absolutely the

BEST in Texas. Fly there and have fun.

Wink (Winkler County - INK)

Airport Cafe

KW Aviation, the FBO at the Winkler County Airport has opened a cafe in the terminal building at INK, the Airport Cafe. It is open Tuesday through Saturday from 9:00 AM until 10:00 PM. Sunday buffet is served from 11:00 AM till 2:00 PM. The Cafe is closed on Monday. The menu has breakfast, lunch and dinner. A very large 27 oz T-bone is offered, with salad, baked potato and hot rolls. Linda Pipkin, the chef, is a master at pies and cakes.

Wink is located 60 miles west of Midland (MAF) and 168 miles east of El Paso (ELP). INK has two runways: 13-31 is 4,500 feet with VOR 13 approach and 04-22 is 3,500 feet. CTAF 123.0. All runways are lighted (PCL). KW Aviation provides a courtesy car.

AMENDMENT

You will be hard-pressed to find a better, more convenient, friendlier place to land and eat than the Airport Cafe, located inside the FBO at Wink, Texas. Charlie and the girls really know how to take care of business. The food is at least four-star, and the Sunday lunch buffet is world-class. They serve steaks after 5:00 PM. On Fridays they have fresh seafood as well. If you're hungry for steak and shrimp, Charlie can take care of you. Next time you're looking for a place to go, give Wink a try. You WON'T be disappointed.

Winnie (Winnie Muni - TX56)

AL T's

Chamber's County is trying to promote this airport and plans to install cheap fuel like Anahuac, according to airport manager Ron Jackson (409-267-8358). They want Trade Days, a mega flea market one mile from the airport on I-10 to compete with Canton. A free limo to the Trade Days is provided by Herbert Thibedeaux, who owns AL T's restaurant and motel. Herbert wants your business and tries very hard to please his customers. The limo driver's name is Bill (409-296-9818) and Ron Jackson recommends we tip him.

AMENDMENT

Al-T's is wonderful Cajun seafood! Herbert Thibedeaux is a very fine owner, manager and person. The food? Just try to finish an "Al-T's Platter." It's so much great food! I especially enjoy the sausage nuggets. Bill, the restaurant manager, will be happy to come to the strip and give you a ride, generally in a very nice limo to Al-T's and then back to the plane. If Bill is off (he does get a day off now and then), Mr. Thibedeaux will drive. He's a really nice man and his restaurant is one of the BEST in the area for Cajun seafood.

There's a pay phone on the pad. You need to take the phone number (409-296-9818) with you because the phone book is missing. Try to give Bill a call beforehand to advise of your ETA. It makes it easier to manage the transportation.

Woodville (Tyler County - 09R)

The Pickett House 

The billing, "Country Cooking served Boarding House Style," is earned by the Pickett House of Woodville, TX. To get there, fly-to the well-maintained but unattended Tyler County Airport. The Pickett people will come and pick you up for the price of a phone call. Seems easy enough. It isn't! In Woodville you dial the number, wait for your party to answer, THEN drop the quarter!

The Pickett House is about a mile away and is more than one building. It is the centerpiece of the Heritage Village Museum. The restaurant itself is yellow clapboard trimmed with white and set off by a rust red, steep, tin roof and a standard 1920s "sit a while" porch. Inside it is vintage "grandma's house." The walls are covered by the Bubba Voss circus poster collection. Some of them date to before the turn of the century.

Lunch is served family style, even if you're traveling alone. Fried chicken, bowls of vegetables and warm, freshly baked bread fill each table. The table trophy goes to the chicken and dumplings and the blueberry cobbler. FIRST RATE! Either is worth the trip.

The price is a wallet-pleasing $6.95 a head.

Bryce Canyon (BCE)

Ruby's Inn

I went to Bryce Canyon (BCE), Utah, for breakfast. The scenery was incredible, and the food at Ruby's Inn was excellent and not too expensive. There is a free shuttle that picks you up at the airport and takes you down the road one mile to Ruby's Inn. The buffet that included cooked-to-order omelets and smoked ham and bacon, sausage, hash browns, waffles, biscuits and gravy, juice, etc., was only $7.00 per person.

Just watch your density altitude, as the field elevation is 7,586 feet.

Green River (Green River Muni. - U34)

Ray's Tavern 

I highly recommend flying to Green River, Utah, getting a car from Redtail Aviation, the FBO, and going into town and having a cheeseburger with fantastic homemade fries at Ray's Tavern. They make a great meal from PBJs to steaks! The great thing is, they make their own fries and do not use some Simplot manufactured potato product. With a burger they give you more fries than you really need. That is how they serve a meal, lots of it, and it has great flavor.

They now have a garden area to the side. So when the sun starts to drop, you can sit out and relax with a good meal.

I recommend you get a couple of friends with planes and make a weekend of sight seeing in the canyon lands.

Kanab City (Kanab Muni. - KNB)

Houston's Trails End Cafe 

The airport is only about a mile from town. Houston's Trails End Cafe in downtown Kanab is super. Most of the old Western movies were shot in this area. You are in the middle of a triangle of the north rim of the Grand Canyon, Bryce Canyon and Zion National Park. Grand Canyon, Bryce and Zion all have airports close. Last time I stopped in Kanab, they loaned me a car to go to town. Very friendly place and spectacularly beautiful.

Ogden (Hinckley Muni. - OGD)

Taildragger Cafe

It is a nice airport and does have some interesting things to do in the area. The Taildragger Cafe is located in the terminal. Usually open for breakfast and lunch, it is the local morning hangout for airport personalities, especially on Saturdays. The food is good, but pretty standard for a cafe. (801-627-1190). There can be a lot to do with three ski resorts within 30 minutes of the airport: Dinosaur Park about eight miles; Hill AFB Aerospace Museum just three miles; Union Station Museum with trains, Browning Arms and automobiles about four miles.

Wendover (Wendover Airport - ENV)

Peppermill Casino

This airport is right on the border of Utah and Nevada. Since Utah does not allow gambling and Nevada does, it attracts a lot of Utahns on the weekends. The casinos offer nice eating and transportation to and from the airport. When I think about the $100 hamburger, this is where we go the most. The Peppermill Casino is a ¾-mile walk. (702-664-2255)

AMENDMENT

The Nevada entry "Wendover NV (ENV)" is actually an airport in Utah, but I see your reason for putting it under Nevada. However, to be complete, you should at least add it to the Utah list as well.

Here it is.

Barre-Montpelier (Ed. Knapp State – MPV)

Sambel's Restaurant 

Sambel's is located on the airport grounds. You can't get more convenient than that. The food is good and the prices reasonable.

Burlington (BTV)

One Flight Up 

Don't forget Burlington, VT's One Flight Up in the main terminal building. It is pretty basic, but you can listen to the planes via headsets and look out the window from some tables. You'll probably park at Innotech, the closest FBO, next to RWY 1/19, and walk the few hundred yards to the main terminal. Innotech closes around 9:00 PM in the summer.

Highgate (Franklin County State Airport - FSO)

Airport Cafe 

There is a lunch counter at Franklin County State Airport in Highgate, Vermont. The menu promises soup, hamburgers and other basic fare. It is operated by Border Air, the FBO. Phone 802-868-2822.

Rutland (Rutland State - RUT)

Amelia's Canteen

I would like to recommend Amelia's Canteen in Rutland, Vermont. Good food, friendly people, and a beautiful place to visit - especially in the fall on a warm day.

AMENDMENT

I'd like to add my praises for Amelia's Canteen at Rutland. I first discovered it this past spring during the 2nd Annual Green Mountain Aerobatic contest. When I take friends for rides, I suggest lunch at Rutland. The food is great, the atmosphere will please anyone into aviation history, and the view from the outside deck is fantastic.

Whistle Stop

While at RUT last weekend in search of the $100 omelet, the friendly staff at Rutland Aviation suggested the Whistle Stop, a converted old train station about 3/4 mile down, the road. An offer of a ride was quickly and graciously accepted, as I had run a race several days prior and my legs had not yet forgiven me for the indiscretion. The old railroad station has been wonderfully restored, with lustrous, homey wood throughout. Green blinds adorn the windows, and the room holds about 12 tables, creating an intimate and casual dining environment. An old caboose sits next door, awaiting eventual restoration into a gift shop. This business, opened about a year ago, is owned and managed by Wanda Webster, who, with her capable and friendly staff, turns out some terrific food!

Breakfast for me was Eggs Benedict, two eggs festooned upon large chunks of homemade bread and delicious, sweet ham slices, amply covered with hollandaise sauce and surrounded by a large pile of home fries, served with as much coffee as we dared to drink. Service was quick, friendly, and genuine. I think I'll defer my annual cholesterol check a few months!

The place is open seven days a week, from sunrise or so through dinner. It offers a wide variety of items, including pizza, chicken breast sandwiches, grinders, seafood dinners, rootbeer floats, and a blackboard full of daily specials.

If you fly in, don't be surprised if they offer you a ride back to the airport. Their telephone number is 802-747-3422. Give it a try. You won't be disappointed!

Vergennes (Basin Harbor - B06)

Airport Restaurant

Basin Harbor, VT, along the east side of Lake Champlain, has a nice little restaurant next to the long grass runway. The approach in the summer is usually over the lake from the north. There are two tall trees at this approach end and many others at the south end. It isn't terribly difficult. This is a beautiful spot to spend the day swimming, eating, relaxing. There are some beautiful accommodations for overnight guests. I highly recommend this spot!

Basye (Sky Bryce - W92)

Resort Cafeteria 

This is a resort with skiing, golf, and other activities. Dining in the cafeteria I would rate one burger. More upscale is Lucio's (not tried). The airport has tie-down space, but no fuel or services. The airport office was closed on the Sunday I visited, thus no Unicom in operation. Bryce is 2,000 feet and is located in a narrow mountain valley. The approach can intimidate the first timer.

What works for me, arriving from the east, is to enter the pattern on a base leg crossing the ridge at 2,500 feet, well north or south, depending on which runway is active. Fly a long final, one could call it a straight-in approach, down the valley to the runway. Departing is just the opposite. Take off and climb down the valley until you have sufficient altitude to cross the ridge.

Lucio's

The review above is pretty right-on with the approach to this interesting airport. It's not listed in most GPS databases, and so you must add it as a user waypoint before you head there. We went on a Sunday, and while the airport has no fuel, there was a nice fellow hanging around the airport who gave us wind directions.

We ate at Lucio's and had a great all-you-can -eat brunch for $7.50. Great views of the ski slopes while you eat. While we didn't have a hamburger, I would give Lucio's a solid four hamburger rating. Service was very nice and friendly! Watch that approach and departure.

Chesterfield (Chesterfield County - FCI)

Airport Cafe 

Chesterfield offers a great restaurant right on the field. Super North Carolina or Virginia-style barbecue with terrific, homemade (not frozen) fries. Friendly waitstaff, low prices, bottomless glasses of iced tea. Perfect place to stop, eat dinner, and change pilots during long, nighttime hoodwork.

AMENDMENT

Chesterfield County, VA, offers a very nice runway (5,000 feet x 100 feet), FBO, and terminal. The terminal restaurant is great - offering a buffet breakfast/lunch. Good eggs, bacon, potatoes, etc., for breakfast and lots of very good barbecue and other foods for lunch. Cost's $6.95 for all you can eat and drink - very good-come hungry. Only problem is that the restaurant is only open 1100-1400 M-F and Sunday. Saturdays by appointment only.

Danville (DAN)

McDonald's

McDonald's is 1/4 mile away (within walking) or take the free courtesy car and leave a buck or so inside the vehicle.

Farmville (Farmville Muni. - FVX)

Prop and Club Grille

Located right on the airport at the FBO is a restaurant that serves lunch. The people are very friendly and the food is delicious and very reasonable. The restaurant is sparkling clean. We ate lunch with a very nice retired lady who flies a C-170 and was just delightful company. The airport identifier is FVX, the restaurant is called the Prop and Club Grille, phone 804-392-9016. Sadly, the people told us that if volume did not pick up, they might close down the restaurant, so call first. It would really be a shame to see such a nice restaurant for aviation close.

AMENDMENT

The Prop and Club Grille is still open as of September 1997! Both the food and service are still first class. I did not get a chance to write down their service hours, so a call before planning your trip might save you from going hungry. There is also a golf course walking distance from the FBO, in case you feel like burning some calories after lunch.

Franklin (FKN)

A.J.'s

The Best Western is right on the field at FKN Franklin, VA. It has a great restaurant in the lobby with buffet and full dinner menus, open seven days a week. It is called A.J.'s, formerly known as Teddy Bears and is located on the south side of the airport 100 feet off the runway. You park your plane in the grass. I fly out of PVG Portsmouth and go there often.

Martinsville (Blue Ridge - MTV)

Windsock

Well, I wouldn't say it's my favorite restaurant, exactly, but it is one you should know about. It's the

Windsock, in the Blue Ridge airport terminal/FBO building. The identifier is MTV. The field is eight miles SW of Martinsville, VA, telephone 703-957-2291. It's mostly sandwiches and light meals, inexpensive and good. They have quite a bit of fly-in lunch traffic on nice weekends. I'm not connected with them in anyway.

Orange (Orange County - W93)

Pizza Shanty 

There is a nice burger dive across the street from Orange Airport. It's home of the "Betty Burger," a sure heart attack on a bun. Only a short 20-foot walk.

AMENDMENT

The Pizza Shanty across from the Orange County, VA, Airport still has the massive Super Betty Burger, a high calorie monster with a full eight ounces of beef.

Saluda (Hummel - W75)

Pilot House 

Saluda has a good restaurant/motel adjacent to the field. It's also a good place for breakfast. The field is next to the Rappahannock River, near the Chesapeake Bay. It's 2,200 feet. Just taxi to the south end of the field and park next to the restaurant.

Echart's

The restaurant across Rt. 3A from Hummel is Echart's, a fine evening dining in the German tradition. Hummel is a progressive airport with a new beacon, state-of-the-art "two-click" runway lights and a completely new and committed "total water free" fuel complex.

Staunton (Shenandoah Valley Reg. - SHD)

Valley Airport Luncheonette 

The airport restaurant has very friendly service. The food is OK, about one step above Mickey D's. The main thing to remember is their hours. Its open weekdays, 8:00 AM until 3:00 PM, and it's CLOSED on weekends. GO FIGURE!

Tangier (Tangier Island - TGI)

Hilda Crockett's Chesapeake House 

I recently visited Tangier Island. There are three restaurants. Two are about two blocks from the airport (nice walk), and the other (Island House) is about a mile. Hilda Crockett's Chesapeake House is family style. That means everybody is at the same table. The food is good and plenty. The other two are more my style and they have the best crab cake sandwiches (five burgers). The bumpy runway is a real treat, and it costs $4.00 to land.

AMENDMENT

Tangier can only be described as one of the best places to visit in a plane. The island sits out in the middle of the Chesapeake Bay surrounded by Pax River restricted airspace. Check to see if it's hot, they bomb things there. The island is small, quaint and easily walkable from end to end. No roads to speak of, just a paved pathway that goes around the island. Hilda Crockett's Chesapeake House is the family-style restaurant, a real treat! Costs about $12.00 or $13.00 and you best be hungry.

Fisherman's Corner 

A short walk from the runway, it has wonderful food and is inexpensive. After your meal, take the walk around the island and meet some of the 700 full-time residents who live their lives relatively secluded from the rest of us. Some still speak with the old English accent of their ancestors. Overnighting is fun with a stay at the Sunset Inn on the south end of the island. Be sure to reserve in advance. The island is truly a destination in itself.

The Waterfront Restaurant 

The Waterfront Restaurant is a nice outside spot next to the dock. Good food, nice "islandy" atmosphere. Inexpensive, a burger, fries and drink goes for $3.25. We abandoned the rest of the group, who sat down at the Island Restaurant. Our reward was bay breezes, conversation with a local boat builder, and a bird's-eye view of watermen fueling up prior to crabbing on the turn of the tide. Friendly black Labs were everywhere!

Wakefield (Wakefield Muni. - AKQ)

The Virginia Diner 

The Virginia Diner in Wakefield is about 40 nm SE of Richmond. It is about one mile from the airport (AKQ) and they will come and get you and take you back, but you may want to walk back to work off some of the extra load. They feature Virginia ham, barbecue, chicken and other regional country dishes, and all the peanuts you can eat. The gift shop sells the hams and peanuts and other things. Good place for breakfast/brunch.

Williamsburg (Williamsburg-Jamestown - JGG)

Charlie's 

They have a real nice place at the Williamsburg-James City County (JGG) Airport. Charlie's serves deli-style rolls and sandwiches as well as the usual burger fare. It's located in the terminal building with tables outside.

The airport is a short drive from Colonial Williamsburg. There you will find a variety of VERY good restaurants suited to every taste and pocketbook. Rental cars, limousines, taxis and even rental bicycles are available. A weekend in this area will not be disappointing!

AMENDMENTS

On a recent instrument cross-country flight, my student and I stopped for a late lunch at JCG on a very cold January day. The lunch was excellent, served in pleasant surroundings. We lingered for some time and were preparing to depart at the end of the afternoon, just as the FBO was closing for the day. We were having difficulty starting our airplane, when two gentlemen from the FBO came out on the ramp and offered to preheat the airplane. They stood out in a stiff, bitterly cold wind for about 15 minutes, preheating our PA-28, which enabled a successful start - and they refused my offer of payment. They could just as easily have left for the day, leaving us on the ramp with a frozen engine and a dying battery. Consideration of that sort should be recognized.

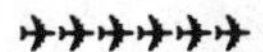

Took a trip to JGG a few weeks ago to look at a plane, which we ended up buying. Food is terrific at Charlie's restaurant. On our first trip down, I had their French dip sandwich - out of this world!! Tender roast beef with melted provolone cheese on their homemade rolls, dipped into beef broth! On our second trip down, I brought some friends along so that we could take our plane home with us. We arrived after the restaurant closed. However, the owner stayed late and fed the hungry bunch of us while we closed on our purchase. Hospitality and food are excellent at JGG. Well worth the trip again!

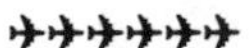

I just received my private pilot ticket and have been busy flying my long-suffering husband all over the state. He's getting better; he only kisses the ramp once when we land now.

Yesterday was a beautiful VFR day. We decided to combine one of my favorite cross-countries with a trip that my husband made one Sunday with my CFI while I was at work. We flew down the James to Norfolk and then back up to Jamestown. We called Charlie's at JGG from Norfolk to make sure they'd still be open when we arrived.

We did barely squeak in on time, and the staff welcomed us like we had all the time in the world. There was a crowd in the little FBO/restaurant, pilots and friends who had flown in, with friends and family who had driven to meet them. There is a nice outside seating area in addition to the indoor dining area. The staff had been waiting for our arrival. There was another plane with pilot and passenger that arrived after us. They radioed in after the restaurant was officially closed and the grill was cleaned, but the cook said to tell them he'd be happy to serve them something from the cold deli menu.

The restaurant area and the FBO are quaint and remind one of a general store or old drugstore soda counter. Both are spotless. The sandwiches are great -- fresh, delicious, and generous portions. The waitress offered to refill our soft drinks before we had a chance to ask. The couple who came in after us were given the same warm welcome. Our total bill for two large sandwiches, Cokes with refills, two servings of pasta salad and chips on the side was $13.00 and change. I have not been to the restaurant on Sundays, but I understand the prime rib special is to fly for.

Arlington (Arlington Muni. - AWO)

Airport Cafe 

This is a nice restaurant at a GREAT airport where they have frequent fly-ins. Good food, good service and good prices, who could ask for more? Hey, isn't this the site of the EAA's annual Northwest fly-in? I think so!

Auburn (Auburn Airport - S50)

McDonald's 

There're a bunch of fast food restaurants just outside the Auburn Airport (S50) (McD's, Jack in the Box, etc.) They are across the street from the south end of the airport. Also, on the east side is a Chinese restaurant.

Bellingham (Bellingham Intl. - BLI)

O'Tool's O'Deli 

The restaurant at BLI, called O'Tool's O'Deli, is located upstairs in the terminal building, with a small snack shop subsidiary downstairs in the departure lounge. There is an excellent view of the runway from the restaurant, and the setting as well as the food is excellent. The whole atmosphere is very aviation-oriented, even to WAC charts set in the glass tables and various aircraft pictures on the walls. The place never seems to be crowded, prices are good, service is immediate, and for those who live in the area I would even recommend driving there for lunch or dinner.

Blaine (Blaine Muni. - WA09)

Burger King 

Blaine is the farthest northwest airport in the lower 48. It sits right at the U.S./Canadian border. If you land on RWY 14, you will turn base and final over the truck customs building. They want you to follow that truck route on final right down to the airport, with a little jog to line up with the runway at the last minute, for noise abatement. On very short final you will notice a Burger King! That's it for eats at this airport, but it's an easy walk. This airport is not for the faint of heart — fairly short (2,100 feet) and narrow, and large trees to swoop over with a seemingly ever-present downdraft if you have to land to the north.

Bremerton (Bremerton National - PWT)

Airport Diner 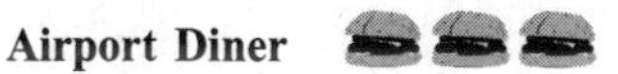

The restaurant at Bremerton (PWT) has reopened. Their menu isn't too bad, and the prices are reasonable. I believe Tuesday night is steak night according to the sign I saw the other day.

AMENDMENT

The present proprietor has long been in the business of selling Fish and Chips around the Kitsap Peninsula and is doing a spectacular job of it. Ron's recipes are tried, true, and good, and even good for you, if you choose broiled recipes!

The restaurant has offered, besides standard fare kinds of breakfast, such spot specials as country ham with grits, red-eye gravy, biscuit and gravy, and "aigs tuh please." This kind of hearty breakfast is usually unknown outside of North or South Carolina, or Georgia! Couple of weeks ago, they offered chicken-fried steak, potatoes, and eggs, which moved a copilot of mine to tears of happiness, it was so good!

Burlington / Mount Vernon (Skagit Regional / Bay View - 75S)

Crosswinds 

A new restaurant recently opened at the Bayview Airport. It is called Crosswinds. The food, atmosphere and service are excellent. I have been there twice in the past two weeks for dinner and plan on more visits. It is located just east of the main terminal building in a new building constructed by the Skagit Port Authority. It has a main dining room on the first floor and a lounge on the second with a superior view of the runway environment. The menu is varied and prices reasonable. If you like good food, this is the one. Parking is in front, with many tie-downs. I have been to many airport restaurants, and this one starts off with a four burger rating.

Cashmere (Cashmere - Dryden - 8S2)

Aplet & Cotlet Candy Factory 

It's just a short walk to downtown - antique shops, coffee and rolls for breakfast, or several restaurants for lunch. The runway, while paved, is a short 1,800 feet, not the place to land a Baron.
The Aplet & Cotlet Candy Factory is there for your sweet tooth. An Aplet is a dried candied apple fruit covered with powdered sugar. A Cotlet is made the same way except it's from an apricot. From Cashmere you can catch a bus to Leavenworth, the ersatz German town.

The Dalles (The Dalles Municipal - DLS)

The Country Kitchen

The city of The Dalles is in Oregon, but the airport is across the Columbia River in Washington. There is a nice little cafe on the field called The Country Kitchen. Be advised, they are open seven days a week, but they close at 2:00 PM.

AMENDMENT

Gisela and Adrienne Sexton just took over the cafe about three weeks ago. They feature daily homemade soup, pies, cakes and specials. Their flame-broiled burger is a definite winner. Their location is in the administration building next to the fuel pumps. The owner, Gisela, told us they are open for breakfast and lunch daily. Also starting this week, they are going to remain open for dinner. Let's all give 'em some support.

Eastsound (Orcas Island - ORS)

Bilbo's

Orcas is one of the San Juan Islands northwest of Seattle. It is actually farther north than Victoria, BC, but is still part of the U.S.

Bilbo's is not right at the airport. There is a 10-15 minute walk from the airport to town, and I would not have known which way to go without a local giving me a ride one way. Bilbo's is a Mexican restaurant with good food and reasonable prices. The portions are not the overwhelming sizes typical of most Mexican restaurants. Be careful when you arrive, as they were open for lunch 11:30-2:30 the Saturday I was there. Call first just to make sure, 360-376-4728.

The biggest attraction in my opinion is the airport itself. It is on the north end of Orcas Island, with water on approaches to either runway. The runway has a noticeable slope downhill to the north, and you end up climbing out low over the water. A lake is just off the south end of the runway. The island is shaped like a big horseshoe, and it's fun to fly between the two fingers. I like the approach from the north over the sound as well. The airport has a variety of aircraft, from an ample supply of Bonanzas to a T-6 to biplanes. When I was there, a couple was camping out of their Tri-Pacer on the grass.

Everett (Snohomish Cnty -Paine Fld. - PAE)

The Sky Lounge

The Sky Lounge is located at suburban Seattle's Snohomish County Airport (Paine Field). It is quite nice, with good service. One of the best parts about this stop is its location. It is a short hop to Boeing Field to see the 747s being built. Tours are available. The Boeing Museum is first class and worth the trip. McChord AFB down in Tacoma also has a great museum. They are kinda' funny about letting us GA types land there. I think they are afraid we'll land inside a Galaxy or something.

AMENDMENT

The Sky Lounge has been open here for years and has good prices. Best of all is the great view of the sunsets and the runway.

The 747, 767, and 777 are made on the north end of the field (not Boeing Field) in the world's largest-volume building, and there are tours given hourly during the week. Well worth taking the time to see.

Friday Harbor (Friday Harbor - FHR)

The Downrigger 

A very popular flying destination in Washington is Friday Harbor, located on San Juan Island. It is a very scenic 45-minute to 1-hour flight (depending on your airplane, of course) from Seattle. The airport is located about 1/4 mile from town, and it is about a 10-minute walk into town from the airport. Friday Harbor is a very popular tourist town, so there are plenty of places to eat. The place I seem to return to when I go up there is located down on the waterfront, just above the marina near the ferry dock. It is named The Downrigger, just to the left (north) of the dock as you approach from town. They have good food. Unfortunately, they raised the prices a bit since I first started going there. It's a pretty decent place though, with an outdoor deck for summer dining.

Just be warned, on a nice, sunny summer weekend the airport could fill up with visiting airplanes. I have never had any problems during the week though.

The Springtree Cafe 

The Springtree Cafe is a relatively short walk from the northeast corner of the Friday Harbor Airport. It is very expensive ($17.00 to $20.00 for the entree) and very exquisite, like a fine French restaurant. Attire is informal, as you might expect in a tourist spot. Every entree except one was a seafood.

Herb's Tavern 

For the best burgers in town, try Herb's Tavern. It's on the corner, one block up from the marina on main street. It is only a 10-minute walk from the airport, all downhill. "Big Bob" is the proprietor; he's always friendly and really appreciates out-of-town guests. After your lunch, stroll downhill to the marina and walk the long main dock for a wonderful time with the folks on the boats. It will be a 20-minute walk back to the airport since it is uphill, but this town is worth it!

Hoquiam (Bowerman - HQM)

Lana's Hangar Cafe 

I have flown to this restaurant 3-4 times. It offers a very complete menu and is reasonably priced. Many people drive from Aberdeen and Hoquiam to eat here.

AMENDMENTS

The restaurant is called Lana's Hangar Cafe. My preference is the B-17, especially on Fridays. This is a sandwich, in case you were wondering, roast beef with melted Swiss cheese, mushrooms and onions. The whole affair can be dipped au jus or in horseradish sauce. It is roast beef on Fridays—all other days it is hamburger. This sandwich is a dream; even now my mouth is watering.

The theme of the restaurant is 50s and includes some nice pictures of WWII airplanes, bombers and fighters. It is a great place to eat. Of course, I'm not biased. Stop by some time and you'll find out what I mean.

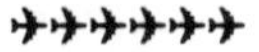

I decided to land at HQM for lunch, and it was a very pleasant experience. The atmosphere is nostalgic - friendly people, exceptional service and excellent food - a clean, well-run establishment. It filled up around noon. More than just aviators stop here. I highly recommend Lana's.

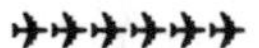

I have worked at Bowerman Field for a little over four years and have always received great food and service at Lana's. I recommend Lana's Hangar Cafe for your consideration.

Ocean Shores (Ocean Shores Airport - W04)

The Homeport 

The airport is really close to the town, a short walk west, and there're a bunch of good places. One of my favorites, last I was there, is The Homeport. Ocean Shores is across the bay from West Port, and like West Port, it is right near the WA Pacific coast.

Packwood (Packwood Muni. - 55S)

Peter's 

The airport is about two blocks off State Highway 12. Packwood has some restaurants along the highway. My favorite is Peter's, which is at the east end of town. Logger-type food is served here. That means OK! Some of these restaurants could be a ½-mile walk, but some are closer.

AMENDMENT

I would just like to add to the blurb for Packwood Muni. It is a wonderful place to fly in. Mountainous terrain surrounds Packwood, and it is located not far from Mt. St. Helen's. Packwood is a great treat to stop into after sightseeing around Mt. St. Helen's.

Port Angeles (Fairchild Intl. - CLM)

Airport Cafeteria 

Fairchild International Airport in Port Angeles, WA, is home to a small GA fleet and some Horizon Air commuter traffic. The airport is well-maintained and easy to find. Watch out for the Horizon guys doing five-mile straight-ins from the east, IFR students shooting approaches, and the 6,000-foot cumulo-granite to the south. The terminal has an okay cafeteria, a gift shop, etc. Definitely a small-town feel: The folks are friendly and the traffic is light. It is a nice stop before continuing west to the Pacific coast or turning back to the more crowded skies over Puget Sound.

Port Orchard (Kitsap Cnty. Airport - 0S8)

Kitsap Cafe

Port Orchard, 0S8, has a nice little Italian restaurant on the north end of the field. There is nothing else there so you can't miss it. You probably won't find it in any directories because it is listed as private. I have talked to the owners of the restaurant. They also own the field, and they welcome any patronage to the restaurant via the airstrip. The runway is paved, but it's a little rough, weeds popping through the tarmac. The quiet location and the price of the food, are worth it, though.

Port Townsend (Jefferson County Intl. Airport - 0S9)

Spruce Goose Cafe

Located just northeast of the Olympic mountains, this airport offers pleasant scenery, interesting airplanes and good food at the Spruce Goose Cafe. The runway used to be grass but was paved a few years ago, all 3,000 feet of it. Lots of parking is available on grass or tarmac. On nice weekends the field draws a wide variety of airplanes that are fun to look at, homebuilts, biplanes, antiques, you name it. A charter operator and fairly new maintenance facility are based there if needed.

AMENDMENT

I regularly visit restaurants at western Washington airports. It's a tough call, but I'd have to vote for the Spruce Goose at Jefferson County Airport as my favorite for good, reasonably priced food; friendly atmosphere; good parking; and a regular stream of aircraft during good weather.

Prosser (Prosser Muni. - S40)

The Red Barn

This hamburger flight comes to a paved 3,400-foot runway, OK even for a Baron. The Red Barn is located on the north side of the field, and there is a parking tie-down just across the road. Breakfast and lunch are good and served at country prices.

Puyallup (Pierce Cnty. Airport - 1S0)

The Hangar Inn

The Hangar Inn is also the regular meeting place for the Northwest Antique Aircraft Association, the second Friday evening, suspended for summer because of flying. This is one of the most active, friendly flying clubs around, whether or not you own an antique. Many members fly their beautiful antiques and classic aircraft to the meeting, too. The restaurant is very accommodating to groups and puts on a really nice meal at a reasonable price.

AMENDMENT

The Hangar Inn at Puyallup is next to my store, Aerocenter. I eat there every day. The deck is great in this February warm weather. They have karaoke on Friday and Saturday nights for those who can stay up that late! The best part is all the flying nuts who show up every day.

Roche Harbor, San Juan Island (Roche Harbor Airport - 9S1)

Roche Harbor Resort 

Roche Harbor Resort is a 5-minute walk west from the airport. This airport is owned and maintained by the resort folks. After landing, taxi to the west end and park in the grassy parking area. There is an honor-system landing fee, $5.00 in the box as you walk off the airport boundary. I'm not really sure what's here off-season. The resort may still be open, but I'm pretty sure the restaurant is not. During the summer, this is a great place to have lunch or dinner, good food and pretty reasonably priced. You can sit in the dining room or if it's nice, downstairs out on the deck. The restaurant is closed November through February. Of all of the airports in the San Juans, this one is the closest to amenities. Furthermore, there is a good kayak rental at the harbor, and you can easily paddle into the Haro Straight, where the best Orca watching in the islands happens. Pretty exciting stuff to be sitting in a small sea kayak with killer whales cavorting about.

Also, at the resort are a small marina, a variety of shops in an old barrel warehouse (or something like that), a little chapel, and a very nice flower garden. Last time I was there, there was a kiosk that looked like they were selling tickets for some kind of bus tour or shuttle or something. If you're there in the evening, keep your eyes open for deer and other wildlife as you walk back to the airport. After sunset, as with all the other San Juan Islands destinations, a clear night presents one with some awesome star-gazing. Services at airport? Zippo! This is a strictly "for getting in and out" airport, operations at your own risk. Fortunately, it's actually maintained pretty well, except the windsock was a little chewed up last time I was there. It's not the best place to be landing after dark. The approach to 24 is from over a hill, with trees, so you're probably better off coming in from over the harbor/resort. Even that approach isn't easy, since the airport is poorly lit and there's no VASI. Also, the runway has a significant slope uphill to the east. Unless the winds strongly oppose you, I'd recommend landing on 06 and departing on 24. Also, be polite. Remember that this is a resort. If your plane can safely climb out with reduced power after liftoff, that's a nice gesture. Otherwise, just remember not to turn before 500 feet AGL. Departures to the east from runway 06 aren't a problem noisewise, but there is that hill with the trees to worry about.

AMENDMENT

Pilots interested in going to Roche Harbor, WA (9S1), should note that it is now a PRIVATE-USE airport; prior permission is required for transient aircraft. The numbers are 360-378-2155 and 378-3500. I haven't tried myself, but I've heard conflicting stories about how easily permission is given. Maybe it depends on who you talk to.

Also, I take issue with the statement, "Of all of the airports in the San Juans, this one is the closest to amenities." I haven't measured it, but I don't think Friday Harbor is any farther from amenities than Roche. Either one is a short walk. They're both great places, with different atmospheres.

Seattle (Boeing Field - BFI)

Museum of Flight Cafe 

There is a new restaurant here. I have heard that it's not as good as the Blue Max, which is closed now, but it's fine for light fare. Besides, the Museum of Flight has its own restaurant. The museum is worth the trip, all by itself.

CAVU Cafe

The new restaurant on Boeing Field is called CAVU. It's OK. It is located in the main terminal building, midfield, on the east side, off the Alpha taxi-way.

AMENDMENT

When first listed, I think the CAVU Cafe (espresso and panini bar) had only been open about a month. Now in full-service mode, they offer a variety of both grilled Italian panini and American-styled stacked deli sandwiches. They also serve great-tasting homemade soups and Caesar or pasta salads, which you can order separately or in combination with sandwich halves. The deli sandwiches, like their "Flyin' Ham," are HUGE! The people at CAVU know the tricks of packaging the "to-fly" orders so they won't dry out upstairs. The newest add to the menu are the "nosewheel" bagel sandwiches, a smaller version of the regular deli types and priced at about $3.00. They serve the full complement of Starbucks coffee and other espresso bar favorites - and fresh pastries too.

The quality and value of the fare at CAVU are absolutely the best on the field, and the Pilots' Store is just next door.

The CAVU doesn't have the great view the Max had. The 1930s plant-filled lobby with its stained-glass skylights is pretty neat in its own way. On nice days there is a picnic table outside right beside the apron if you must watch the action while you eat.

Shelton (Lake Nahwahzel Resort)

Lake Nahwahzel Resort

Lake Nahwahzel Resort near Shelton, WA, has excellent food, featuring prime rib on Saturday evenings. Reservations are required for the prime rib feast. It is the best prime rib I have ever eaten...anywhere! The resort is located on the south shore of this smallish lake and has a very good seaplane dock that will accommodate several seaplanes. They have a full breakfast, lunch, and dinner menu.

Shelton (Spencer Lake Tavern)

Spencer Lake Tavern

Spencer Lake Tavern on Spencer Lake near Shelton, WA, has the best burgers in Mason County. It's floatplane access only, dock available for tie-up.

Snohomish (Harvey Field - S43)

Harvey Field Cafe

Near Arlington, this is one of WA's many very-active general aviation airports. Sit outside in good weather and eat your lunch or dinner while watching the skydivers drop in. Excellent chocolate shakes (no malts though...sigh...).

Tacoma Narrows (Tacoma Narrows Airport - TIW)

Rotor Bar and Grill 

The restaurant at TIW is now open again after remodeling. It's now called the Rotor Bar and Grill. It has an expanded kitchen and bar. The menu has a bunch of new items on it as well as some of the old fare. This place was always a favorite for me and it's back, better than before.

AMENDMENTS

I was forced down here due to weather the other day, and was warmly greeted at the newly opened Rotor Bar and Grill restaurant. Well-decorated and good food, I recommend it.

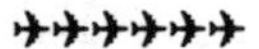

Just want to second the previous comments made about Rotors. The decor is nice, the food is great, the service is friendly, and the price is right (especially on my college student/student pilot budget). In addition, its location allows an unobstructed view of the entire airport and the sunset (provided no aircraft are parked in the tie-down spots about five feet from the entrance). I took my fiancee there when we weren't flying; it's that good. Last time I was there, I watched a student pilot on her first solo.

Vancouver (Evergreen Field - 59S)

Applebee's 

There are lots of restaurants within walking distance of Evergreen Field. Applebee's is my pick. Evergreen is a great airport, and lodging is nearby at the Phoenix Inn. Lots of shopping around, too. This is my favorite place. A great place to take the spouse; she/he can shop or eat while you hang around the airport and talk to the owner, Wally Olson, who has taught flying for over 55 years.

Walla Walla (Walla Walla Cnty. - ALW)

Airport Cafe 

I fly to Walla Walla occasionally and have always found a good meal and friendly help. The restaurant is in the airport terminal.

AMENDMENT

I had lunch at this restaurant for the sixth or seventh time in the last five years. The food continues to be good, the prices reasonable, and the service prompt and courteous.

West Port (West Port Airport - 14S)

King's Royal Restaurant 

Walk along the main road, about a ½ mile northwest from the airport, and you'll find King's Royal Restaurant, pretty good Pacific fare. Extra bonus: There's a nice waterfront state park just across the street.

AMENDMENT

West Port is a great place to go if you're into deep-sea fishing, whale watching, or any other sea charter activities. There's a waterfront district about ¾ mile northwest of the airport. Walk to the right on the main road that passes the airport parking lot. There's a good selection of gift shops, candy stores, restaurants, motels, and boat charter establishments. The waterfront area is compact and self-contained, so transportation is really not needed for a weekend of fishing. However, you can rent bicycles at a go-cart track that you'll pass going into town.

There's an entrance to a state park about halfway to the waterfront area. The distance to the state park is about the same as to the waterfront. The park is for day use only.

At the northernmost part of town there's an observation tower overlooking the entrance to Gray's Harbor. The view can be spectacular when the waves pound the breakwater.

Woodland (Woodland State - WA71)

The Oak Tree 

About ¼-mile walk from the tie-down at Woodland, WA, are several good eating establishments, including The Oak Tree.

Yakima (Yakima Air Terminal - YKM)

The Restaurant at the Airport 

I have been flying from Pasco, WA, to Yakima for dinner with friends since 1977, and the restaurant located upstairs over the main terminal building has always been a four-star dining experience. The waiters are as attentive as the food is outstanding, and the view from our booth of runways 9-27 (usually around sunset) keeps us "airplane junkies" challenged to pay as much attention to our lady friends as we do the traffic. After all that, the late-night Bonanza flight back to the Tri-Cities over the lower Yakima valley makes a great capper to the evening.

AMENDMENT

The girlfriend and I went to The Restaurant at the Airport for brunch last Sunday and found the experience to be mixed. The service was somewhat bizarre, with long waits for coffee, even after the waitress had commented that we needed more, and the food was definitely nothing special. We recommend avoiding the breakfast sausage. However, the views and location couldn't be better, and they validate parking for up to three hours for you.

The price was average, with $10.00 per person for the all-you-can-eat brunch being the going rate. It's probably cheaper for children. As usual, sitting around and watching the traffic more than made up for any shortcomings, and we'd still recommend the place.

The restaurant is open daily, with the kitchen closing at 2230. Opening times are 0730 on week-days.

Davis (Windwood Fly-In Resort Airport - WV62)

Amelia's

Windwood Airport near Cannan Valley, WV, has an excellent restaurant located on the airport. The name is Amelia's, and there is a lounge called Yeager's Bar and Grill. The airport is the only fly-in resort in WV. There are home lots for sale on the airport. The resort is located in a four seasons area: snow skiing, hunting, fishing, golfing, whitewater, hiking, and lots of outdoor activities. Very nice place for a stop!

Huntington (Tri-State Airport - HTS)

Tri-State Airport Restaurant

Try it, you'll like it. This is a great place to grab a meal. They feature burgers, chicken, salads, etc. The food is very good, tastefully prepared and presented. Service is very good, also. The restaurant is in the main terminal, an easy walk from the FBO.

Lewisburg (Greenbrier Valley - LWB)

The Greenbrier Hotel

For those with bizjets and pocketbooks to match, the Greenbrier Hotel is located in White Sulfur Springs about 12 miles from the airport. They provide limousine service. Bring you golf clubs and your American Express Card, the Platinum one!

Bones Diner

Bones Diner is undoubtedly the epicenter of fine dining in the western hemisphere.The epitome of western culture and theology. All of mankind is equal at Bones Diner and on busy nights, they, too, shall stand and wait. It is visited by kings, scholars and desperate men.

The Lewisburg, West Virginia, airport serves Bones Diner in White Sulphur Springs. It is about 12 miles east of Lewisburg off of Interstate 64. Bones Diner is on Main Street at the intersection of Big Draft Road, right down the street from the Greenbrier Resort. Late at night on weekends it is crowded with patrons from the nearby clubs. Bones Diner is a unique dining experience.

Martinsburg (Eastern WV Regional - MRB)

The Airport Cafe

The Airport Cafe on the southeast side of the Martinsburg, WV (MRB), airport has the finest (read greasiest) hamburgers and freshest everything else around. Best of all, from a poor aviator's standpoint, the prices are low, since the owners grow many of the ingredients themselves. Typical lunch for two is about $8.00 - $9.00.

Morgantown (Morgantown Muni. - MGW)

Airport Cafe

Morgantown, WV, has a great airport with the friendliest tower I've ever encountered. They also have a great place to eat on the field in the terminal building. You taxi right up to the front door, and the guys from the FBO will even put out the red carpet for you as you exit the plane, even my '62 Cherokee 160. Prices are reasonable, the food is good, and the dessert menu is out of sight. If you go on a Saturday in the fall, you can even try to catch some great Mountaineer football over at West Virginia University. The food gets a good four and the football gets a solid ten.

Parkersburg (Wood County - PKB)

Helen's

There's a great restaurant in the terminal building, adjacent to the GA ramp, excellent food, reasonable prices, and a good view of the field. Apparently, the owner is into radio-controlled models, as there are plenty suspended from the ceiling.

Rainelle (Rainelle - Private)

Airport Cafe

Rainelle Airport is a 3,800-foot grass strip on top of a mountain in Rainelle, WV. Meals are offered Friday, Saturday and Sunday from April to October. The food is great, and Squire Haines, the owner, is likely to meet you as you deplane. Inside, you will be greeted by Pansy and Karen, who will feed you West Virginia style. A great place and good folks. Don't miss it. The Rainelle Airport is five miles from the Rainelle VOR on the 120 degree radial.

The strip is EXTREMELY smooth and has excellent approaches in both directions. During the summer months, the mountain top is always a lot cooler than elsewhere and thus is always a pleasure to visit. Permission to camp is usually granted by the operator, Squire Haines, when asked. I find the food, the clean, cool air, and the friendliness of the people absolutely refreshing.

Appleton (Outagamie County - ATW)

Victoria's

While this particular restaurant isn't right on the field, you can access it via a hotel shuttle at the public terminal. Known as Victoria's, this Italian restaurant consistently rates "outstanding" among locals and visitors alike. It's quite easy to spend less than $10.00 - $12.00 per person for a dinner that few visitors have been able to finish. While quantity is often used to make up for lack of quality, this isn't the case here.

ATW is THE alternative airport for visitors to the annual EAA convention and is actually the best place to land when visiting - less air traffic, friendly tower operators and lots of room for tie-downs.

Baraboo (Wisconsin Dells - DLL)

Ho-Chunk Casino

If you dare, fly into Baraboo-Wisconsin Dells. The Ho-Chunk Casino is little more than a stone's throw from the airport. The casino will pick you up in their courtesy car, so you can, if you're lucky, enjoy an inexpensive and expansive buffet. Rental cars are usually available at the airport. Nearby Wisconsin Dells is filled with large motels and hotels, complete with indoor and outdoor waterparks and a multitude of restaurants. Baraboo is charming, with its Circus World Museum.

Brodhead (Brodhead Field - C37)

The Sandpiper

If you ever want to experience a little aviation Valhalla, you must attend the annual Pietenpol Exposition at Brodhead, WI. This grassroots fly-in highlighting the active resurgence of Bernie Pietenpol's Model-A-powered parasol occurs in this sleepy little southern Wisconsin town every year during the first weekend of Oshkosh. Yes, they know what they're doing. They vote on this every year, and every year they decide to have it this weekend. It actually works out really well for those types of flyers — low and slow and proud of it. They intentionally skip the big-weekend at OSH.

The airport has three runways, all grass, but the crossed runways are closed during the event. Camping is encouraged since Speth's Swiss Chalet (only motel for miles) fills rapidly.

The fly-in puts on several eating events. In the mornings, the Lion's Club hosts pancake breakfasts. There is usually some type of lunch. Friday night there is a fish fry, and Saturday night there is a brats and corn roast. This is fantastic food; the corn comes right out of the field you pass over on short final!

There are lulls in this flying and eating orgy. So when you get a hankering, beg/borrow a ride to the Sandpiper Cafe. There is a clunker airport car, and the airport itself is owned by a very active EAA chapter, so there are lots of rides available.

The Sandpiper is a basic diner with, get this, a tea-room attached! The diner has good burgers and very friendly service. They, too, have good corn. The tea-room is a notch up in fancy. The French onion soup is good, as are the basic entrees. Neither the diner nor the tea-room will set you back much, $5.00 - $10.00 for a good lunch.

Note that The Sandpiper is popular but small. That is how we ended up in the tea-room. The diner was full—the first time we ate there.

I give it a 3 out of 5 burgers. It is basic fare, but well-done. The cafe would make 3.5 to 4, but for some reason, the folks who run the diner feel overwhelmed by the fly-in crowd, so the service isn't as friendly as it could be. Just a small quibble, though.

Delavan (Lake Lawn Lodge - C59)

Lake Lawn Lodge 

Try Lake Lawn Lodge in Delavan, about 30 miles west of Lake Michigan and 10 miles north of the WI/IL state line.

The Lodge owns the paved runway and provides transportation across the street and through the parking lot to the restaurant. It is on the north end of Lake Delavan. When approaching from the east, it is difficult to locate. The Lodge itself has boating, an outdoor pool, horseback riding, golf, etc. The food is good, great Sunday buffet!

AMENDMENT

The people at both the airport and lodge are really friendly and accommodating, and the breakfast buffet at the Lodge ($6.95) was excellent. It would be a perfect weekend spot, with a great-looking golf course and many other recreational activities right there.

Eau Claire (Eau Claire County - EAU)

Connell's II

Connell's II is located right in the terminal building at Eau Claire, WI. It is a nice, clean, usually busy restaurant that caters to locals as well as fly-in people. Portions are huge! I sometimes order a ham steak and have enough left to take home for another entire meal. Prices are very reasonable and quality is consistently good.

Fond Du Lac (Fond du Lac Cnty. - FLD)

Schreiner's Restaurant 

Fond du Lac, at the southern end of Lake Winnebago, is home to Schreiner's Restaurant, a 50+ year old eatery with some of the finest food you'll find - at an affordable price. Best of all, the owner is a pilot and goes out of his way to bring us in!

Land at Fond du Lac County Airport (runways 09/27 and 18/36 are both paved), and the friendly folks at the FBO, Fond du Lac Skyport, will call Schreiner's for you. Within minutes they'll have a courtesy car there for you, free of charge.

They serve all types of food, from breakfast all day to steaks. We fed our family of two adults and two kids for less than $20.00, and walked away stuffed.

When you're done, simply tell the cashier that you need a ride back to the airport and away you go! Terrific service, terrific food, and they bend over more than backwards for pilots!

AMENDMENTS

We at Schreiner's just celebrated our 59th anniversary. The owner that is a pilot, Bernard Schreiner, retired five years ago. I have worked for the company 28 years, having started as a dishwasher in high school. My wife and I purchased the restaurant from Mr. Schreiner in June 1992.

We CAN'T offer the shuttle service during EAA and IAC conventions, but a local service is available during that time at nominal cost. Walking is nearly two miles one way, and if folks walked to the restaurant during AirShow week, we'd be too busy to take them back. Sorry!.

N/S runway at Fond du Lac (FLD) is undergoing a complete rebuilding; it should be operational by the Airshow. N/S landings have been on the taxiway during construction! Contact them at 920-922-6000 for info and an update.

The Fond du Lac area code changed to 920 on July 26, 1997.

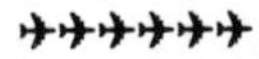

Thanks to the extraordinary hospitality of Paul Cunningham, I was able to sample his restaurant's fare while on the grounds of the Oshkosh EAA Convention. Paul provided lunch for the Warbird group (read huge) and invited me to join in.

He served the BEST meatloaf I have tasted. It was truly my loss that I couldn't work in a stop at Fond du Lac this year. In 1998, I won't miss it!

Muslim's go to Mecca; $100 Hamburger fans MUST go to Schreiner's!

Fort Atkinson (61C)

Lou's Riverview

Friendly faces, fun atmosphere, and great sandwiches are available at Lou's, a four burger rating! Closed on Monday's. Exit runway 3, taxi north on the grass, park before the last group of hangars, and walk about one city block.

Janesville (Rock County - JVL)

Recently, I did a long cross country. One stop on my tour was Janesville, WI (JVL), in southern Wisconsin. I stopped in at the terminal's CAVU Cafe for lunch. Here's the report: Parking is great! The main wall of the cafe faces the intersecting runways and is basically all window. You can even park your plane in front of the windows so everyone you're dining with will understand why you can only afford to purchase a cheeseburger and fries for lunch. Not that anything's wrong with the CAVU Cafe cheeseburger. I opted for the smaller, 1/3-lb. burger (with fries) for just $2.95! The burger they're famous for is the 1/2 lb. of ground chuck for a buck more. It also looked as though this was a local hot spot - the restaurant was filling quickly as I left at noon. If you're in southern Wisconsin, don't miss it! Great food, service and view!

AMENDMENT

Great little restaurant! Usually lots of interesting and unusual airplanes, too!

Unfortunately, the FAA ramp check inspectors like the unusual planes and the good restaurant there, too. They were out in force a few weekends ago.

La Crosse (La Crosse Muni. - LSE)

Amelia's

Amelia's restaurant is in the terminal building.

Lake Geneva (Americana - C02)

Grand Geneva Resort

Grand Geneva Resort is the former Playboy Club, located in Lake Geneva. This resort has a working ski hill, golf courses, and a beautiful restaurant overlooking the very scenic Lake Geneva area. We visited for their breakfast buffet, which, at around $10.00 per person, was very reasonable for what we got. We're not talking scrambled eggs here; nope, they had eggs benedict, salmon — the whole nine yards. Served on real china, with cloth napkins, no less. This is not your greasy spoon FBO.

The FBO shuttles you to and from the restaurant and will gladly fill you in on what else there is to do in the area. Until the new sectionals come out, your best bet is dial in Lake Lawn Lodge (C59) coming out of Burlington (C52). You will fly directly over Grand Geneva Resort.

AMENDMENT

The resort is excellent - three restaurants, two golf courses, a ski hill. A shuttle will pick you up from the airport. Food and service are really first rate.

The airport has no identifier yet, but it is located two miles northeast of Lake Geneva, Lat 42 36'53" Long 88 23' 27". Unicom frequency is 122.8. Runways are 5/23 3,545 feet x 75 feet right traffic for runway 5. Pattern altitude is 1,835 feet MSL. 100LL is available 24 hours. The telephone number is 414 -248-8811, Ext. 3812.

Madison (Truax Field - MSN)

Jet Room

One of my favorite places to fly-and-eat is actually at my home field. The Jet Room at MSN isn't pretentious. It's just an airport lunch counter with a few tables. The food is good, the portions are generous, and the prices are very reasonable. The Twin Engine Burger is my personal favorite. Unlike the typical greasy spoon airport restaurant, you can order anything on the menu and get good food. The restaurant is actually inside the FBO building on the east ramp, so you get a good view of what's happening out there.

MSN is the center of Class C airspace. The controllers are some of the best around, courteous and friendly. The FBO is one that prizes service. Ask the controllers for a city tour on your way in or out. The view of the Madison isthmus, capitol, and university area is a favorite.

Manitowoc (Manitowoc Cnty. - MTW)

Plane View Restaurant 

The Plane View Restaurant in Manitowoc, WI, is just across the street from the airport at end of runway 17. Good food. Breakfast, lunch and dinner. 414-682-1001 Pilots fly in from many locations. Magnus Aviation (414-682-0043) has an excellent FBO 5,002 feet x 100 feet RWY.

Milwaukee (Timmerman Field - MWC)

Skyroom 

The Skyroom at Milwaukee's, Timmerman Field (MWC) is open for lunch and dinner. It is open every day but Sunday between the hours of 11:30 AM and 3:00 PM for lunch and 4:00 to 10:00 PM for dinner. Taxi right up to the restaurant, good food and a view overlooking the ramp. 414-461-5850. I rate it four hamburgers for convenience and food.

Just a short distance (about two blocks) from the ramp, there is another restaraunt called Maxium's. You get a nice view of the airport while having breakfast, lunch or dinner. Open seven days a week, with a varied menu and moderate prices.

Monroe (Monroe - C33)

Baumgartner's 

I would like to recommend the city of Monroe, WI. Baumgartner's restaurant is known across the country. Getting to the restaurant is a simple matter of calling for a taxi (cost was $4.00 for three people). Baumgartner's is representative of German restaurants, with lots of beer and sausage. They have coffee for the pilots!

The town is friendly, the food is good, and the price is quite reasonable. Going on a Sunday, however, one will find most of the town closed up. This is best visited on a VFR Saturday.

Oshkosh (Wittman Field - OSH)

Friar Tuck's

At Oshkosh's Wittman Field (OSH), ask to taxi to northwest parking and walk through the gate there to go to Friar Tuck's. This place is EXCELLENT, with great sandwiches and a really neat atmosphere. Friendly people too!

Winemaker

About a mile from the terminal building at Oshkosh, there is a restaurant called the Winemaker. They have a 32-oz. prime rib that is unbelievably good and BIG. I get at least one, sometimes two every year I make the fly-in. You get to wait a while, even with reservations. The salad bar is excellent also, and I can also speak well of the ribs, but I would fly (from Texas) to Oshkosh for the prime rib, even if they canceled the fly-in. Actually, I have never had bad food in Oshkosh anywhere. They have at least two Mexican food places that are excellent and you can get DOS XX beer up there, too. (Note to Texans — ask them to bring the real hot sauce out from the back, otherwise it's much too mild for our tastes.)

Prarie du Chein (Prarie du Chein Muni. - PDC)

The Black Angus

There's a place up in Prarie du Chein, Wisconsin, called the Black Angus. It's a few hundred yards down the road from the airport, but they had a mean fish fry on Friday nights the last time I was up there.

Reedsburg (Reedsburg Muni. - C35)

Longley's Restaurant

Now, Reedsburg is hardly your haute-cuisine center of the world. On the other hand, it has a nice little GA airport, a good airport cat, and it sits smack in the middle of hamburger heaven. There are several fast-food restaurants within easy walking distance of the ramp. Culver's Frozen Custard has excellent billion-calorie burgers, in addition to all the dessert prospects. There's also Longley's Restaurant, which is a standard Midwest family eatery with competent food, but really excellent pies.

Shawano (Shawano Muni. - 3W0)

Launching Pad Bar/Restaurant

The Launching Pad Bar/Restaurant, directly across the street from the airport office in Shawano, WI, has GREAT burgers, fries, fish, etc. You will get change back from a $10.00 bill every time. One paved and two grass runways.

Siren (Burnett Cnty. - RZN)

Dana's Family Restaurant 

Dana's Family Restaurant, is a two-block walk south of the runway 31 threshold. It has a gift shop, art shop, and a museum. The tavern north of the restaurant has house-smoked pork and beef. Great ribs!

Superior (Bong - SUW)

Belknap's South 

Belknap's South is a fine restaurant just off the threshold of 13. Taxi to and park in the rear of the building. They have picture windows to watch airport activities, including skydivers landing next to the restaurant and packing their chutes on the lawn.

Voyager Village (Voyager Village - Y05)

Voyager Village Restaurant 

There is a small airport in northern WI called Voyager Village. The runway is in the middle of a golf course and is paved 3,500 feet. The restaurant is an 8-iron away from the ramp, and it smells like heaven when you deplane. There is a full restaurant and bar. Eat steak if you go. Also play golf, a very nice course. I recommend going in the summer, since the field is private, but anyone can land. It has no services. See the Green Bay sectional. Stop at Siren Airport for all services before heading over. The restaurant keeps shorter hours in winter months. It's a good idea to call ahead: Restaurant: 715-259-3382; Airport Info: 715-259-3910.

Watertown (Watertown Muni. - RYV)

The Steak Fyre 

We stopped at Watertown overnight on our way to the big show at OSH. We were allowed to camp out overnight on the field. There are three places to eat within about a 300 yard walk of the airport. Two of the restaurants are fast food type. The people at the FBO recommended that we try The Steak Fyre restaurant. We had an excellent meal at a very reasonable price. You can choose your steak and cook it yourself or they will cook it for you. This is a very good stop prior to joining the line of traffic going into OSH.

Cheyenne (Cheyenne - CYS)

Owl Inn

My favorite restaurant in southern Wyoming is the Owl Inn in Cheyenne. You fly into the Cheyenne Airport (CYS), tie down at the FBO and then walk two blocks west. They serve breakfast, lunch and dinner. The food is very good and the prices are very reasonable. One of my family's favorite stops.

AMENDMENT

This is a great place. We flew up to CYS on Saturday evening and visited the Owl Inn. Food was great, large portions, reasonably priced! I talked to the owners and they say they do not get a lot of business from CYS. A nice walk, even at night. Sky Harbor does have a courtesy car for your use.

Medicine Bow Brewpub

Last weekend, we discovered a great place called the Medicine Bow Brewpub, which just opened earlier this year. Tie down at Sky Harbor FBO. They have a courtesy car. I make it a practice to always buy fuel at an FBO if I intend to borrow the car. Drive into Cheyenne on Central, Highway 85. Turn left at 17th Street downtown and the restaurant is on the right-hand side. The prices are very reasonable, and the menu is very eclectic. The pheasant quesadilla appetizer is great. Also, they have homemade rootbeer with complimentary refills. Of course, I was not able to sample the hand-crafted beer, but one of my passengers reports that it is excellent. Very high recommendation! I may even DRIVE back to Cheyenne some time!

Pinedale (PNA)

Della Rose

The FBO had an old Cadillac for a courtesy car, which we took the several miles into town for breakfast. He had recommended Della Rose, which turned out to be great! It is about midtown on the west side of the main street. You can't miss it. Food is great, reasonable cost, and the atmosphere is Old West all the way. The scenery around Pinedale is incredible. We flew around the Wind River Range and over dozens of clear blue lakes. The scenery alone was worth the trip.

Riverton (Riverton Regional - RIW)

Airport Cafe

At Riverton, Wyoming, on the airport, at the commercial terminal building, not the FBO, is the Airport Cafe. It is so good and reasonably priced the locals drive out from town for Sunday dinner! Mom bakes the pies, and there were 30 of them available! (307-856-2838)

Saratoga (Shively Field - SAA)

Mom's Cafe

Mom's Cafe is in beautiful downtown Saratoga. It is only a mile from the airport. After lunch take a six-block walk to Hobo Hot Springs for a dip!

Thermopolis (Thermopolis Muni. - THP)

Legion Supper Club

At the Thermopolis Airport? Yes, exactly across the parking lot is the Legion Supper Club. Great food, prices not cheap, but reasonable for the value received. I had the NY strip steak and it was GREAT! Sunday breakfast buffet is available from 9:00 AM until 1:00 PM for $5.75 + tax. Legion Supper Club opens 5:00 PM for dinner, no lunch. (307-864-3918)

Thermopolis, Wyoming Airport is up on the ridge, northwest of town. The airport managers are wonderful local folks. You know you're at a real airport when the airport dog is as friendly as "Gin," who's last known cohort was "Tonic," named for a previously favored drink of the owners.

Part 2

The Rest of the World

Isla Martin Garcia

Bife Con Papas Fritas 

Located north-northeast of Buenos Aires, flying to Martin Garcia makes you enjoy a nice sight of the big city, the Rio de la Plata and the Parana Delta. Not so much for the two or three fine places where you can get the famous $100 hamburger, or better, a classic "bife con papas fritas" (steak with french fries). The island is a wonderful place to walk around, then find a quiet place by the river to rest under the fresh shadow of a "sauce" tree and maybe to read some good stuff. Spring or autumn weekdays recommended or later than 5:00 PM on weekend days to avoid high temperatures and crowds.

Pueblo Lobos (Pueblo Lobos)

LV Bar 

We live in Buenos Aires, Argentina, and we fly in a Cessna 172. Every weekend in Pueblo Lobos, we make a combination of parachute jumps, golf and swimming in the pool. After that, we take a great meal of asada criollo y ensalada mixtal at the LV Bar.

Pueblo Lobos is about 120 km from the capital of Buenos Aires.

Melbourne (Moorabbin Airport)

Wardy's

Wardy's is right on the airfield. It is a small cafe with everything from the standard Australian meat pie and tomato sauce to full-cooked meals or a quick cappuccino. Famous for their milk shakes and steak sandwiches. Of course, you can get a real hamburger there, too, plus some healthy salads. Wardy's is the unofficial meeting place for aerobatic pilots.

Sydney (Sydney Intl.)

McDonald's

If you allow McDonald's to rate as a restaurant, then here's a fun way to drive through, fly through, that is. Sydney International has a McDonald's restaurant in the Ansett Terminal.

It is possible for GA aircraft to land and take off at Sydney, but beware, during peak periods there is a hefty landing fee (AUD$250). However, outside these times the fee is a reasonable $25.00. After landing, taxi to the GA apron, tie-down and walk to the Ansett Terminal.

This makes it an interesting trip to get some "Mackas." Last time I flew in a C150, I had a B747 and two B767s at the holding point waiting for me to land.

Order your food takeaway and walk to the end of the Terminal and you have commanding views of the three runways.

Tirol (St. Johann)

Airport Guest House 

A real nice airfield with authentic Austrian food is St. Johann in Tirol, Austria. The small restaurant is found right on the field. As you process (actually the Austrian border official does it for you) your international flight plan into/from e.g., Germany, order one of the vanilla-strudels. You're gonna' love it! Mind that the airport is closed for departure during lunchtime. With the above meal option, it should not create too much of a problem. Landing fees are moderate for Europe, but it's better to not fly into Innsbruck. The landing fee at this place might spoil your day (>$50.00 for C182). So try St. Johann for about $18.00.

Andros

Lighthouse Marina 

The only place I have eaten on Andros is Lighthouse Marina at Fresh Creek. Nice airport, but no fuel. Food was OK after a $20.00 cab ride.

Bimini (S. Bimini Intl.)

The Complete Angler 

On Bimini in the Bahamas there is a restaurant at the airport on South Bimini. Of course, there is also Gary Hart's favorite place. The Complete Angler, where Hemingway used to frequent. To go to The Complete Angler, you must take a cab from the airport on South Bimini to a ferry that takes you to North Bimini, aka Alice Town. This is a FUN stop!

Opal's

The best place to eat in Alice Town is Opal's, up the hill across from the Bimini Big Game Club. She has three tables in the front room of her house. The food is good. I suggest the cracked conch. While you are never really sure what the charges are, they seem fair!

Casa Grande 

From the Bimini airport take the cab and then the ferry over the beautiful clear aquamarine water to North Bimini ($5.00). Upon disembarking, walk (25 minutes) or rent a golf cart ($20.00/hour) and head for the Casa Granda at the northern tip of the island for lunch and a swim. The view is magnificent and the food good. A BLT was $5.00 and came complete with fruit and pickle. The pace in the islands is very slow, so do not expect fast food service. The Casa Grande is the only three-story building at the end of the island. From the beach, you realize that it is shaped like a boat! After the meal, which is served in a small area overlooking the coast, take a swim in the warm water or try some snorkeling. You will probably have the long white sandy beach all to yourself! The locals need to spend time cleaning up some of the refuse on the beach. Overall, you probably won't notice it due to the beautiful scenery and gorgeous ocean.

Crooked Island, Pittstown Point

Airport Restaurant 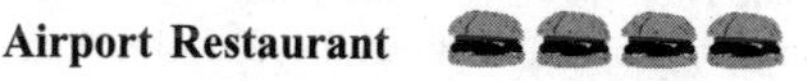

You should add Pittstown Point, Crooked Island, Bahamas. I realize it may not be a close reach from Florida or Nassau, but it is the Family Islands and you will be treated royally! Pittstown has a 2,000-foot paved runway and you are welcome to stop. Contact Cindy Bates at 1-809-344-2507.

AMENDMENT

The greatest place to go is Pittstown on Crooked Island in the Bahamas. The restaurant and the rooms are only 100 feet from the runway. Cindy Bates, the innkeeper, and her staff do a wonderful job. The food is superior, if you like grouper and lobster twice a day! There are other choices for those few who don't like fresh seafood. Rooms are $95-$115 per night. The runway is 2,000 feet of asphalt and in pretty good condition. For those who feel faint of heart on only 2,000 feet, I have seen all sorts of twins in there: Barons, Senecas, Aztecs, Skymasters and even an Aerostar. Springtime brings the tradewinds. They are strong but usually down the runway.

To quote Cindy, "When you get to Pittstown, you find there is nothing to do, and when you are ready to leave, you find you are only half done."

Eleuthera (North Eleuthera Airport)

Angela's 

A very short car taxi to the water taxi to Harbour Island and pink beaches. I suggest lunch at Angela's and conch salad.

Eleuthera (Governor's Harbour Airport)

Unique Village 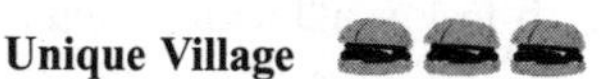

The next airport on Eleuthera is Governor's Harbour. The most wonderful lobster thermador was at Unique Village. There is a local hangout pizza place that has conch pizza. Both of these are at Palmetto Point, just south of Governor's Harbour. They also now have a movie theater this year.

Great Exuma

Two Turtles Inn 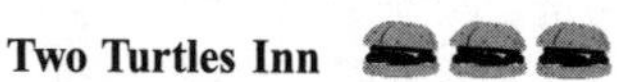

Two Turtles Inn on Great Exuma, George Town, has a wonderful barbecue Friday nights. Well worth a stopover.

Great Harbour Cay (Private)

Tiki Hut

The easiest and best place to get to from the States is Great Harbour Cay in The Berry Islands, approximately 30-45 minutes past Bimini almost due east. There is an absolutely gorgeous beach across the highway from the airport and the best hamburger in the Bahamas at the Tiki Hut there. All is within easy walking distance of the airport, which is private but serves as a port of entry and welcomes tourists. The 4,500-foot runway is in fair shape, but it's average for the Bahamas. No fuel and a $5.00 landing fee.

Harbour Island (Harbour Island Airport)

The Landing

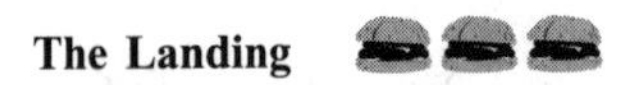

There is a wonderful European restaurant in Harbour Island a short five-minute trip from ELH called The Landing. A superb French chef prepares true French/Italian food. One of the best in the Bahamas!

Marsh Harbour

Wally's

Marsh Harbour has the cheapest fuel in the Bahamas at Zig Zag. The $1.85 beats $3.85 the last time we were at Stella Maris. Anyway, Marsh Harbour has the only working stoplight in the Out Islands and a number of excellent restaurants. Try Wally's or Mango's.

San Salvador

Riding Rock Inn

A little farther out is San Salvador and the Riding Rock Inn. Radio on your approach and the Inn will send a car for you. It has the best conch chowder in the world, and I used to live in the Florida Keys so I do know a little bit about conch. Great wide runway but no fuel, you'll need to stop at Exuma or Stella Maris. Stella Maris is also a good lunch stop.

The Three Ships

The Three Ships restaurant in Cockburn Town, San Salvador, is great for more than hamburgers. Faith serves the best cracked conch, pigeon peas and rice, and grouper you have ever had, and a mean cold Kalik!

AMENDMENT

Those flying into San Sal will see a lot of construction equipment at the east end of the runway. The runway is presently being lengthened and upgraded, from 4,500 feet to about 7,600 feet to accommodate jets. Time of completion is late 1997. The restaurants, Three Ships in Cockburn Town and Riding Rock Inn, are still the best on the island.

Treasure Cay

Chris' Beach Hut 

Down the Abacos is Treasure Cay. A $7.00 per person taxi ride takes you to Chris' Beach Hut for the best barbecue conch burgers around. It is next to one of the 10 best beaches in the world. For $35.00 you can drink and eat all day long! You could take the water taxi over to Green Turtle Cay. There you'll find several places to eat as well as a really neat little town.

Walkers Cay

Marina Restaurant 

The northern Bahamas have Walkers Cay. The restaurant by the marina has better-than-average cracked conch. The runway starts and ends at the water and is 2,500 feet. Fuel is available if you are a hotel guest. Everything on the small island is within walking distance.

AMENDMENT

I'd have to up the rating for Walker's Cay restaurant to at least four burgers if not five. It's expensive, but I thought it was fantastic!

Ostend

Fisherman's Quay 

Really close to the English Coast and an alternative to France is Ostend. At weekends the seafood stalls are spread all along the Visserskaj (fisherman's quay). You can buy cups of prawns, mussels, or the Belgian national dish, fries with mayonnaise! On the other side of the road are dozens of really good, inexpensive restaurants selling seafood, steaks and continental specialties. To get into town, regular bus routes from the airport to town or taxis are the thing. The area is extremely attractive, and all the shops are open on Sundays for inexpensive wine to load up your aircraft. Check your weight and balance!

Abbotsford, British Columbia (Abbotsford Airport – YXX)

H.F. Wings Restaurant and Pub

The Red Baron closed and in its place H.F. Wings Restaurant and Pub opened. You can taxi right up to the doors.

Boundary Bay, British Columbia (Boundary Bay Airport – CZBB)

Skyhawk

Boundary Bay Airport is Canada's 5th busiest airport and BC's busiest general aviation airport. The restaurant, Skyhawk, has a killer south view of the runway system and the large apron area, along with a mean choice of burgers, sandwiches, and a bacon and egg breakfast. I must admit, it's selection of Chinese dishes is my attraction. Rudy does all the cooking himself and always has a daily special in the $5.00 - $7.00 range. It's located in the main terminal of the airport. There are transient parking spots in front of the terminal that are well marked with a "P." There is also parking to the West on busy days. The restaurant has an outdoor deck for lazy summer afternoons. It has CUSTOMS, which makes it easy for flights from the United States, which is only five nautical miles to the south.

The airspace is Class D, with an operating tower on 118.10 from 7:00 AM to 11:00 PM (Pacific time). There are two 3,800-foot hard-surface runways. It is VFR almost every day and very easy to see with its big white WWII hangar glaring out at you.

With eight flight training schools, there is always a flurry of action at Boundary Bay CZBB.

Braeburn, Yukon (Braeburn Lodge - EK2)

Braeburn Lodge

About 60 miles north of Whitehorse, Yukon, is Braeburn Lodge (EK2). They make the biggest cinnamon buns on the planet. The strip has a sign next to it designating it as the "Cinnamon Bun Strip." All other items on the menu are huge too, and delicious. A common stop for local aviators and the usual destination for most student pilots completing their first solo x-country.

Calgary, Alberta (Calgary Intl. Airport)

The Spruce Goose Cafe

The Spruce Goose Cafe is located at Calgary International Airport in Calgary, Alberta. Make your way to the Shell Aerocentre ramp. The SGC is at their FBO. It's not a cafe, despite the name, it's a very upscale restaurant. The prices are rather high but not outrageous, and it has service and food to match. Well worth a visit. I have gone there for a couple of special functions. You can sit by the window and watch the planes while you eat. There are several other facilities in the same building, including a flight planning room with telephones.

Calgary, Alberta (Springbank Airport - CYBW)

Wings Bed & Breakfast

The Wings Bed & Breakfast at Springbank in Calgary is not a restaurant, but a B&B operation located on the airport property, directly across from the Esso FBO (Southern Aero Aviation) 403-286-6151. Springbank is about 15 minutes west of Calgary, elevation 3,937 feet MSL. It's a GA feeder airport built to relieve GA pressure on Calgary International Airport.

AMENDMENTS

The Wings B&B at Springbank Airport is also a fully operational restaurant. It opens at 7:30 AM each morning. Stop by, the food is good.

✈✈✈✈✈✈✈

This place has fabulous food, especially the charcoal-barbecued steak sandwiches on the weekends. As a pilot, I have come to appreciate this restaurant. Unfortunately, I am developing a serious C of G problem due to all of the great food.

The Springbank Airport Restaurant

The Springbank Airport Restaurant serves excellent food for breakfast and lunch. It is located in the Calgary Flying Club building just south of the tower and FSS buildings and east of runway 34. You can sit and watch the activity on the runway while you enjoy the plentiful home-cooked food. 100LL fuel and tie-downs are available at the club as well.

AMENDMENT

The Springbank Airport Restaurant is open in the Calgary Flying Club building from 0730 through 1400 daily and closed during bad weather. It is often referred to as the Calgary Flying Club Cafe. The restaurant is a favored hangout for club members and visiting pilots and is often filled with spirited aviation-related discussions.

The food is standard lunch-counter fare. Short order-staples are supplemented with daily breakfast and lunch specials that add a "home-cooked" flavor. The food is tasty on average and occasionally downright delicious. The prices are right and the staff is courteous.

Large windows look out onto the Club's ramp, runway 16-34, and a panoramic view of the Rocky Mountains.

Fuel, friendly advice, and other aviation amenities are available at the Flying Club. At present there are no ramp fees for aircraft stopping to visit during the day. The telephone number is 403-286-8443.

Campbell, British Columbia (Campbell River Airport)

Airport Cafe 

The small hamburger cafe at Campbell River Airport is nice and friendly. This is a great town to visit, with picturesque countryside covered with forests. It's a seaplane base and is close to Comox RCAF base, which has a world-class airshow every OTHER year.

Cheltenham, Ontario (Cheltenham Airport)

Brampton Flying Club 

Brampton Flying Club in Cheltenham, Ontario, is north of Toronto. This is a very good restaurant and good facilities. Worth the stop!

Chilliwack, British Columbia (Chilliwack Airport)

Airport Cafe

Chilliwack Airport is about 60 miles east of Vancouver, BC. The restaurant here has excellent food and nice, cozy atmosphere. It is located in the terminal building. You can sit outside and eat in the summertime. A VERY quiet place to get away from the hustle and bustle. It is similar to the restaurant at the Qualicum Beach Airport.

AMENDMENTS

The restaurant in Chilliwack, BC, is deemed by the 99s and all other hungry pilots to have the best pie in the province. Even the controllers at the other airports understand the meaning of a "pie run to Chilliwack."

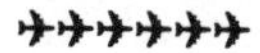

The restaurant at Chilliwack Airport has to have the BEST pies in existence! They're made fresh every morning, and they've got more flavous than Baskin Robbins. The Mud Pie is my favorite, and it's huge.

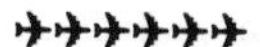

Chilliwack Airport is approximately nine miles east of Abbotsford and has a paved runway. ATF is

122.7. Fuel and maintenance are available. It is famous for the pie. People have been known to travel 300+ miles on a Sunday to have a piece. It should also be noted that Chilliwack is the home of Murphy Aviation (Rebel, Renegade, Maverick).

Courtenay, British Columbia (Courtenay Air Park)

Betty's Place Cafe

I just flew in to Courtenay Air Park, which is also a floatplane base, and had breakfast at 2:00 PM! Behind the big three-bladed prop is Betty's Place Caf-eh (Yeah, Canadian, eh?). A great spot for a hamburger, but they also serve up ham n' eggs or any other breakfast at any time. All kinds of burgers and sandwiches plus three kinds of coffee are available. The sign on the door says, "We're open 'til we close." I suppose that means that as long as there are customers, they are there to serve you. The cafe was taken over last September by Bev Archer, a wonderful cook and conversationalist. Next time you're around Vancouver Island, give it a try. You won't be disappointed.

Delta, British Columbia (Delta Air Park)

Delta Air Park

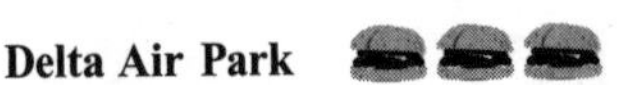

Delta Air Park is a sort of an EAA place. Lots of interesting general aviation airplanes base here including Texans (Harvards) and WWI replicas. Several helicopters base here as well. The coffee shop cum hamburger joint is not bad, but it is not always open, especially in the winter.

AMENDMENT

We just about lost this airpark. The property changed hands and is now managed, under license, by local pilots and enthusiasts in cooperation with the GVRD (Greater Vancouver Regional District) Parks Board and the provincial government. We hope you enjoy your stay with us and ask you to respect the airfield rules and regulations. They are designed for your safety and the benefit of flyers and our neighbors.

Com. 123.3
RWY 07/27 Turf 2500 feet (old X wind RWY 12 closed)
Services: 100LL, 80/87
Procedure: Remain east of noise-sensitive area. (Farm west of RWY 25)
Circuits 600' and No low 'n overs
Coffee shop is now on the honor system to support the cause.

Fort Chimo, PQ (CYVP)

Kuujjuaq Inn

Kuujjuaq or Fort Chimo, PQ, is located along the Koksoak River. Other than the river the tundra is featureless. The main paved mile-long runway parallels the river. The half-mile gravel runway slopes up and away from the river. Most summer transportation in this frontier town is by unlicensed ATV four-wheelers. A pizza restaurant and a hotel (Kuujjuaq Inn 819-964-2272) located about 300 yards north, is a nice walk. Once a year a supply ship comes in August. Beer and other necessities are flown in once a month, at which time several hundred thirsty natives stand in line at the Co-op.

The Inuit natives took control and renamed the towns. This is why they have two names.

Frobisher, Baffin Island, Northwest Territories (CYFB)

Toonoonik Hotel & Restaurant 

A "place of many fishes" or Frobisher, Baffin Island, NWT. Here's the lowest cost av gas in Canada, $1.90 per gallon and a two-mile-long runway with VOR & DME. From Houston this is halfway to Europe. This outpost has very good accommodations for having a supply ship once a year in August.

Florence and Roger Melanson operate the Toonoonik Hotel & Restaurant (819-979-6733; fax 4210), about two blocks north of the airport. It is the best value in town. The restaurant has a beautiful view of the bay. It has 45-foot tide changes. They have great grilled fish and for breakfast large fresh pancakes with Canadian maple syrup. Roger said immersion suits were available here.

The senior Customs Inspector, Germain Sacirette (819-979-6714; fax 2857; pager 4844), was very pleasant to work with, unlike the U.S. type. He gave us "Arctic Circle Explorer" certificates. By chance, we met again for breakfast at the Toonoonik and ate together.

There is a community center where natives have folk dances, several schools, a mall with a theater and an indoor public pool. Twenty hours of summer daylight gives you plenty of time. In winter the northern lights are reportedly spectacular, but the locals say it is very very bad luck to whistle at the northern lights.

Goderich, Ontario (Goderich Ontario Airport - CYGD)

Airport Cafe 

Goderich, Ontario airport, CYGD, has two restaurants. One is at one end of runway 28 and the other at the 10 end. The 28 end has special parking for the restaurant on the highway, made from an old runway extension. The other restaurant is near the terminal building. The place on the road is your "truck stop" type, and the one at the terminal is good for lunch.

Guelph, Ontario (Guelph Air Park - NC4)

Aviation Cafe 

Guelph Air Park is a private airport with two paved runways 14-32 2,400 feet and 05-23 2,100 feet. The restaurant is open 7:30 AM - 2:00 PM seven days a week for breakfast and lunch. Try the Breakie, or if you're really hungry, the full breakfast.

Now a story. A C195 pilot out of Hamilton was complaining about his $100 hamburger as he was placing his order for hamburger. The restaurant owner (it was separately run back then) quickly shot back, "Well, you didn't spend it here!"

Haines Junction, Yukon (CYHT)

The Raven 

Located on the edge of Kluane National Park, 71 nautical miles west of Whitehorse, Haines Junction is one of the prettiest places in the whole Yukon (and there are plenty of those). Floatplanes can put down a short way from the airstrip on Pine Lake. Either way, it's a good 30-45 minute walk to town along the Alaska Highway, nothing someone who can pass a pilot's medical can't handle. BTW, people here pick up hitchhikers. If they don't stop, they're probably tourists.

The place to take your appetite is The Raven. This is no ordinary restaurant...definitely NOT a greasy spoon. Be prepared for fine dining with a European flair. People travel for hundreds of miles to eat there. Trust me. Oh, and bring your wallet too.

If it's breakfast you are looking for, you should go to the Mountain Cafe. This place appears to be just another one of those gas station restaurants. Try the Eggs Blackstone. If you think you've had good eggs before, prepare to be reborn.

Tucked away next to the Kluane Park Interpretive Center (a must-see) is a little bakery that makes a great variety of delicious breads, pastries, soups and sandwiches. It's worth the trip.

Some info on flying in: Watch the cable span on Paint Mountain (next to Pine Lake). The strip (5,000 feet gravel) has a bit of a rise in the middle, but nothing too significant. You can pretty much follow the highway right from Whitehorse.

Hanover, Ontario (Hanover/Saugeen Municipal Airport - PN4)

Airport Cafe

Great place for breakfast, especially Sunday morning when the community comes out to watch all the crazy pilots come in. Excellent breakfast (especially the local sausage) and lunch selections. It is located in the beautiful southern Ontario farmland.

Hamilton, Ontario (Hamilton Airport)

Canadian Warplane Heritage Museum 

The new Canadian Warplane Heritage Museum on Hamilton Airport is open daily, with specials on Thursday and Saturday. You can view the museum, then eat! They have all the regular airport grub and a wonderful steak dinner on Thursday for about six bucks.

Kingston, Ontario (CYGK)

The Captain's Corner

The Captain's Corner serves a varied menu. Call up on 123.35 and Lana will have it waiting for you. I spend $100 per month there. Food's yummy and you get a window seat 100 yards short of the button to runway 01.

La Grande Riviere, PQ (CYLG)

The Radisson Restaurant 

The Radisson Restaurant has good French food for about $12.00 each person. Stay at the Carrefour Hotel $55 double. This is on the edge of tundra; cars can go no further north. Buildings are on pilings because of permafrost. You can take a free one-day tour of one of the world's largest hydro-electric projects here. The air is very clean. Good fishing is reported near here.

Langley, British Columbia (Langley Airport)

Ikarus

Langley is about 30 miles east of Vancouver and is a small, somewhat hard-to-find field (keep your eye out for the large unique church near the field). Ikarus is a great Greek restaurant right on the field at the north end. Good food, decent prices, and on a nice day you can sit outside and watch the airport activity. If you fly in, dessert is free with dinner.

Linden, Alberta (Linden)

Mennonite Restaurant 

The best fly-in lunch in south central Alberta, Canada, is at Linden, approximately 40 miles north by northeast of Calgary, Alberta (south of Three Hills, the nearest published airport). Linden is a very small town, with two good restaurants and a historic site that was used for hunting by Indians. Pilots requiring 5,000 feet of paved runway need not bother. The grass strip is on the very eastern edge of town, about 2,500 feet long with a power line on the southern threshold. Take off to the north unless a strong southerly wind is blowing. The strip is flat, with no unusual challenges, but can easily blend into the surrounding crop. A pilot who has confidence on grass strips will have no trouble with a light load in a 172, 182, etc. Beware of marginal performing planes with a takeoff towards the south (powerlines). After parking your plane beside the old folks home, walk 300 feet to the local Mennonite restaurant or down the main drag to the local Chinese place. The food is excellent and service is friendly. Best of all, there are no tie-down fees or long waits for clearance. Use the Three Hills radio frequency of 123.2 for local traffic. No fuel; no winter maintenance; look out for gopher holes.

Lindsay, Ontario (NF4)

Airport Restaurant 

Lindsay has a very good restaurant. Despite the size of the airfield, it is quite busy, since it borders on a highway. Particularly good (and large) breakfasts. The homemade butter tarts are very popular. Worth the trip!

Midland, Ontario (Midland Airport)

World Famous Dock Lunch 

Sorry folks, I'm not a flyer, but like Jimmy Buffet, I like my cheeseburger in Paradise. Midland, Ontario, has a nice airport with access to several great burger joints. The World Famous Dock Lunch is floatplane accessible. Try it!

Montreal (Montreal)

Le Bifsteak

I am a lineman for the Dorval Shell Aeorcentre in Montreal, and I have driven thousands of pilots to Le Bifsteak. It's a huge steakhouse, not expensive but super excellent. It takes four minutes to drive there.

Nanaimo, British Columbia (Cassidy)

Connections Cafe

Nanaimo has a great restaurant in the main terminal building called Connections Cafe. Scott Forrest, the owner, claims he makes "the best darned hamburger in the world" and after trying out his hamburger and fries, I have to agree with him! The fries are also especially delicious, better than I've ever found anywhere else. Sandwiches include roast beef, ham, turkey. Scott makes a mean, hot, spicy chicken Caesar salad for those with gourmet taste. The cafe is open from 0600 to 2000 PST every day, and is also licensed. This restaurant is definitely worth a visit, and for taste alone the $100 hamburger is cheap!

Oliver, British Columbia (AU3)

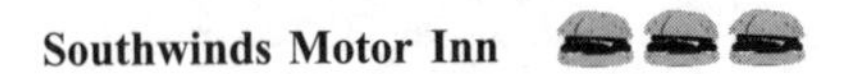

Southwinds Motor Inn

Oliver, BC, Canada is airport AU3 in the Canadian Flight Supplement. It is 3,200 feet x 50 feet asphalt, just north of the Canada/U.S. border. At the south end of the strip is quite a nice motel with a real nice restaurant. You can taxi right up to the motel.

Orillia, Ontario (Mara Airport)

Runway 28 Cafe

There is a good flying restaurant at the Mara Airport called Runway 28 Cafe. It is located about 10 nm east of Orillia, Ontario.

Oshawa, Ontario (Oshawa Municipal Airport - YOO)

Crosswinds Cafe

Oshawa Municipal Airport is on 400 acres in the center of town. Which is home of General Motors in Canada. YYZ is 35 miles to the west. YPQ is 45 miles to northeast. Oshawa has a control tower and handles corporate jets. An alternative to Buttonville is Markham, Ontario, which is 20 miles northwest of Oshawa. Field constructed 1941 for pilot training. Two on-field museums display military tanks and an air collection, including a Tiger Moth, Harvard, DC 3. Lake Ontario is two miles away for charter fishing. Ice fishing is 45 minutes north. The best kept secret in Canada.

The Crosswinds Cafe is located upstairs in Hangar #3, the new hangar on the south side of the airport. It's home of the Oshawa Flying Club. Big changes have been made, with a new air terminal on the north side of field, new tie-downs and T-hangers. Drop in for good food and good connections to the heart of Toronto.

Ottawa, Ontario (OFC)

Ottawa Flying Club

One of Canada's original flying clubs, the Ottawa Flying Club, Ontario, Canada, does well at what a club is supposed to do, and the cafe is part of this club. As a fly-in destination OFC is a good location for a lunch visit or an overnight. The friendly staff go out of the way for visitors and club members alike.

Nearby is the famous National Aeronautical collection, and as the nation's capital, Ottawa has lots to offer.

Parry Sound, Ontario (Georgian Bay Airport)

Mark's

The airport is located just south of Parry Sound, Ontario, on the west side of Hwy. 69. It is not very big. If you locate Muskoka Airport, which is larger, and look to the W/SW, you should be able to locate it.

Great food just off the airport at either Mark's restaurant or Memories of Muskoka.

Pender Island, British Columbia (Pender Island - Private)

Browning Marina and Pub

Pender Island is across Georgia Strait between Saturna and Main Island and behind Samuel Island. A one-way-in-only grass strip. Good uphill rollout. The strip looks short from the air. Don't worry, I have seen twins go in and out of there. Watch for downdrafts. Touch down past the helipad. It is a private strip and you are totally on your own. Service is a zip! Walk down the runway to the road and head south. On your right you find a minimall. Keep going and take the next left and you end up at the Browning Marina and Pub. Excellent food is only a 10-minute walk from your plane.

Peterborough, Ontario (CYPQ)

Cockpit Cafe

The airport is at N44 13 48, W78 21 48. It has a hard-surfaced runway, 09/27, 5,000 feet x 100 feet, and a turf runway, 13/31, 1,788 feet x 50 feet. It has taxis, fuel, maintenance and customs (limited hours).

The Cockpit Cafe is in the terminal building and is open seven days a week from 0800 to 1600. It has recently undergone a change in ownership, and the new operator produces an excellent bacon/egg/toast/coffee breakfast for $5.00.

Qualicum Beach, British Columbia (Qualicum Beach Airport)

Airport Cafe

Qualicum Beach Airport is about 35 miles north of Nanaimo, BC. The restaurant here has excellent food and nice, cozy atmosphere. It is located in the terminal building. You can sit outside and eat in the summertime. A VERY quiet place to get away from the hustle and bustle. It is similar to the restaurant at the Chilliwack Airport.

Red Lake, Ontario (Red Lake)

Lakeview Restaurant

The Lakeview Restaurant in Red Lake, Ontario, 51 04 N by 93 48 W (approx.) is a good spot for a meal for float flyers. The food is plain but good and plentiful. Prices are reasonable and service is quick and friendly. A section is licensed. Opens early and stays open till late evening.

It is located on Howey Bay on Red Lake and is right in downtown Red Lake. You can land your floatplane right in the bay and tie up at Greens or Red Lake Airways. An excellent maintenance facility is located across the bay. This is floatflyers' country, and you'll never feel more at home than in Red Lake. For a real thrill, stop by in July for the annual Norseman Festival.

St. Catherine's, Ontario (Niagara District Airport)

St. Catherine's Flying Club

Niagara District Airport is located near St. Catherine's, Niagara Falls, Ontario, Canada. It is virtually across the river from Niagara Falls, New York. Niagara Falls International Airport is about seven miles away.

St. Catherine's Flying Club is located on the airport property. This social club fires up the barbecue every Saturday from noon to 4:00 PM during flying season. The French Fly and Roadhouse Burger for a fiver are proving very popular with the students in groundschool and fly-in guests.

The proximity to Niagara Falls, Lake Ontario and upstate New York make this a popular fly-to destination. The airport staff try hard to please the customers who arrive by aircraft.

This is a typical midsize airport, with taxis, maintenance, fuel and customs all available. The NDB approach is adequate, as the WX rarely goes to 200 limits. There are three runways available, a 5,000-foot 06/24 and two secondaries of 2,500 feet 01/19 and 27/09.

Timmins, Ontario (CYTS)

The Fish Bowl Restaurant

The Shell fuel server, Steve Feather (Manager ph. 705-267-4884; fax 268-6472), has a courtesy car. (Customs ph. 705-264-8503.) It is 10 miles to The Fish Bowl Restaurant in Timmins on the Mattagami River, Hwy. 101. Jeans and kids are OK, but it is elegant. There is a gold mine tour and a minor league baseball field on the same main road through Timmins.

Cochrane has a train to Moosonee on James Bay, 20-foot tides. Train, airplane or boat is the only way to this town.

Toronto, Ontario (Toronto City Center - CYTZ)

The Lady Luck Casino 

Hamburgers are pretty good; fries are the best; real "homefries" style made from real potatoes, skin still on the ends! Also good fresh sandwiches, muffins, bagels, etc. Lunch is the best time, breakfast is not as good.

Vancouver, British Columbia (Floatplane Aerodrome)

Flying Beaver Pub 

Directly to the south of Vancouver International is a small aerodrome for floatplanes, with the world-famous Flying Beaver Pub. This is a classic pub with some of the best food on the lower mainland. On a nice summer day you can sit on the patio on the Frasier River and watch the sun go down over the Pacific.

Waterloo, Ontario (Waterloo - Guelph Regional - CYKF)

Airport Cafe 

A new restaurant facility has opended at Ontario, Canada, at the Waterloo-Wellington, also known as Waterloo-Guelph Regional Airport, lat. N43-27-32; long. 080-23-05. Excellent food and service.

Whitehorse, Yukon (CYXY)

Claire's 

Whitehorse, the capital city of the Yukon Territory, sees a lot of travelers in the summertime. Those who choose to fly the Alaska Highway on their way through to or from Alaska will find Whitehorse a pleasant place to spend a few days.

Accommodations are available right next to the Apron if you can't bear to let your plane from your sight. For a very reasonable fee, you can stay in one of the clean well-lit rooms above Aero Tern Fuels. Contact Roger O'Brian at 403- 667-2622.

On the airport there is a restaurant in the terminal building called Claire's. This may be the only airport restaurant in the world that people eat at because the food is good, not because they have to. Also, the high perch and glass walls give an excellent view of the goings on at the airport.

Just across the highway from the airport, try The Airport Chalet for reasonably priced, good hearty fare and friendly service. This is where the Mounties eat, so it must be good.

Downtown (a five-minute cab ride) I recommend The Cellar. This restaurant in the Edgewater Hotel specializes in prime rib and they have it right. Just down the block, the Eden restaurant offers a varied menu but specializes in Vietnamese cuisine. If you do nothing else in Whitehorse, try Vietnamese combination plate #2 here. The Blackstone Cafe roasts its own coffee beans and serves an excellent soup and sandwich. Try the No Pop Sandwich Shop they serve no pop, thus the name.

This is the Yukon's most widely recommended restaurant. If you just happen to be in Whitehorse on a Friday, then go to the High Country Inn for the seafood buffet. The best Sunday brunch in town is offered at the downtown Westmark Hotel.

Wyoming, Ontario (Sky View)

Sky View 

Sky View is at N42.59, W82.08 and is about 10 miles east of Sarnia, Ont., and 3 miles north of Wyoming, Ont., next to Highway 7. It is a grass strip, and they have maybe 25 to 30 planes hangared there. The restaurant is open seven days a week, and there is also a gas pump with avgas. Truckers and local farmers along with pilots make the restaurant a popular place to eat. The menu is your eggs and bacon, hamburgers, hot roast beef sandwich, always served with a smile. The restaurant seats about 30 to 35. Sunday mornings are busy with pilots flying in for breakfast.

Alderney

The Rose and Crown Hotel 

With France all around, the Channel Islands are sometimes forgotten. They do not belong to the UK and the status is non-EEC. A bit like the relationship between Puerto Rico and the U.S....but different.

There are three Channel Islands with airports, Jersey, Guernsey and Alderney. Alderney is the smallest of the three and has a good airport with air traffic control inside the Channel Islands Control Zone. The Rose and Crown Hotel is a wonderful old stone building run by a South African from Cape Town. He provides an extensive menu and wine list. Both have, as you can imagine, a South African flavor and plenty of channel seafood...lobsters, moules mariniere and the like. The island is not subject to UK tax or VAT, and therefore fuel is only 37p/Litre, which is less than half that of mainland Britain. It's easily accessible for a day trip from the UK but better as an overnighter to really gorge yourself on the great food. I met a French pilot who had flown to Alderney, SOLELY to gas up, because of the price! In addition, as if it wasn't enough, the Rose and Crown does a "Pilot's Package" for overnights, which includes a free landing (normally Sterling 8.50), free parking for up to 72 hours, free use of a car (true!) weather on request. A PPL accompanied by an IR-rated pilot need only pay Sterling 10/night for the safety pilot, and of course, flight plan filing by Fax. Of course, the wonderful Duty Free prices are worth it all on their own. Currently whiskey is under Sterling 6/bottle, gin and vodka are cheaper. Tel: UK 01481 823414 or Fax 01481 823615. Write to the Rose and Crown, Alderney Channel Islands, for a brochure. Check carefully your procedures for flight in the Channel Islands Control Zone. A wonderful place!

Bernay, Normandy

Airport Cafe

I just went to Bernay in Normandy, home of Mudry Aircraft. The food was great at the outdoor cafe/bar. I can't remember the name of the place, but it was good. It's the only one at the airport.

Mers Le Treport

La Plage

A lovely little channel town with dozens of wonderful restaurants along La Plage (the seafront). All sell inexpensive, gorgeous French food with good wine on tables set for watching the world go by.

The little airport at Le Treport is on top of the hill above the town and is PPR (prior permission required). A quick phone call to 33 35 865 634 is courteous. Paradise on a sunny summers day! Great for impressing the new girlfriend.

Le Touqet, Pas De Calais

Airport Cafe Le Touqet 

The restaurant on the field at Le Touqet in the Pas De Calais area of France is extremely popular with pilots from England and France. It's 40 minutes from the Southern English airfields.

It is 100 meters from the AIS office at LTQ and is very busy with locals eating on weekends. A good sign! Their moule marinieres (mussels in white wine) is outstanding, and almost anything on the menu is to be recommended. There's a wine list, as you would expect from a French restaurant, but nightstop if you decide to sample!

The nearby town of Le Touqet is packed full of good restaurants, but if you want to visit for lunch only, the excellent Hotel Moderne is to be recommended.

Aalen-Elchingen (Flugplatz Aalen)

Flugplatz-Restaurant 

Aalen-Elchingen is a nice GA airfield located on the Schwaebische Alp, about 80 km east of Stuttgart. The restaurant there offers very friendly service, excellent cake, and good dinner, all at very reasonable prices. On weekends in summer things can get a little crowded, both on the tarmac and in the restaurant, as the place is well known.

Braunschweig Airport (EDVE)

Volare 

Check out Volare, the nice Italian restaurant at my home base. They have a very interesting menu. You can just have a coffee and some cake in the afternoon or a complete dinner in the evenings. I would award four burgers to this restaurant, although fortunately, you can't get any burgers there (I guess).

Dortmund (Flughafen Dortmund - EDLW)

Flughafen Restaurant 

There is one small airline, Eurowings, flying from Dortmund to many destinations in Europe: Munich, Berlin, London, Paris and others. The rest of the traffic is general aviation. The Flughafen Restaurant is in the airport terminal building. I don't know the menu at this time because I haven't been there for a while and it changes often.

Idar-Oberstein-Göttschied (Flugplatz Göttschied)

Flugplatz-Gaststätte Göttschied 

I would like to tell you about my favorite German fly-in restaurant. The menu is quite simple, the food is fantastic and the view is second to none. The towns of Idar and Oberstein are small German villages located between Mainz/Wiesbaden and Trier. The airport itself is located on top of a rather large hill, making for some interesting takeoffs and landings. It is a grass strip and is host to a variety of aircraft, from ultralights to motor gliders to the largest high performance singles. The restaurant is located right on the airport.

— Menu —
Sierra - Schweinekamm saftig gebraten
Lima - Eine ganze Schweinelende, zart und mager
Romeo - Das Roastbeef für Kenner
Foxtrott - Zartes Rinderfilet

You are encouraged to radio in your order on 122.85. Just use the designator listed above.

Closed on Mondays and holidays, otherwise open from 11:00 AM to 11:00 PM.

Koblenz (Flughafen Koblenz)

Airport Cafe 

The airport at Koblenz is the home of a Greek restaurant which is located in the main terminal building. The restaurant has a very full menu, Greek and German, so there is something for everyone, and a complimentary after-dinner drink comes with all selections (pilots use caution!). In the summer, it is absolutely delightful to sit on the terrace and watch the traffic. Watching glider activity is a favorite pasttime of many locals. Check it out.

Paderborn-Lippstadt (Paderborn-Lippstadt)

Rietberger Aeroclub 

Paderborn-Lippstadt, a tower-controlled field with more than 7,000 feet of runway (06/24), is located in the northern section of Germany. It is halfway between the big cities of Cologne and Hanover and is rather inexpensive for landing, i.e., a single-engine landing will cost about DM 20 or less, depending on weight and possible noise certificate. A real advantage is their opening hours, from 6:00 AM to 10:00 PM in the summertime. One of the two flying clubs (Rietberger Aeroclub) runs a very nice restaurant adjacent to the fuel pumps. It is open seven days a week and offers reasonable prices and good food. Should a foreign pilot need any help, just ask for one of the chairmen (Mr. Lipsewers or Mr. Gieffers) or any member of the flying club. A list of hotel accommodations is available at the GA terminal.

Rothenburg/Tauber Germany (Flughafen Rothenburg)

Rothenburg Biergarten 

Rothenburg is the GA airfield right next to the medieval city of Rothenburg. The restaurant is open every day, serves basic, but really tasty food and features a nice “biergarten,” with a playground for kids.

Stuttgart (Echterdingen - EDDS)

Hotel-Restaurant Lamm 

The Hotel-Restaurant Lamm is positioned in Germany near Stuttgart Airport (Echterdingen - EDDS). The address, Hauptstrasse 98, Echterdingen.

It is about three minutes away from the airport by car. Take a cab. It will cost about DM 10. The hotel rooms are very comfortable. The boss is a Greek with a friendly, relaxed manner. He's always in for a good joke. The food is of high quality, and there are several meals from a Greek, German, or French menu.

The cook, named Charley, is the best one in all of Germany, so get ready for hot meals!

So everybody who's tired of smelling gas from planes, get into that place and be spoiled by the nice hosts of Hotel Lamm in Echterdingen. Prices are quite acceptable!

Nuuk (BGGH)

Cafe Rudolph

Cafe Rudolph is located on the waterfront in Nuuk (or Godthab, the capital of Greenland) in an old boat repair shop above "Santa's North Pole Workshop." There is a beautiful view of the fiord, with occasional whales and icebergs. The $10.00 grilled salmon is very good. A bus leaves the airport every hour on the hour weekdays 7:00 AM-9:00 AM and 3:00 PM-5:00 PM. Cost is $1.00 and it services the entire town. There are two hotels in this town of 15,000 people with a 12-year waiting list for permanent housing. The Seaman's Center Hotel's (fax 22104) special price for pilots is $100 cash US per night for a room with two single beds and a shower and a good breakfast included. The main "Hotel Gronland" will cut their $250 price in half if you say you are going to stay at the Seaman's Center.

Due to prevailing high pressure in central Greenland gusty south winds are common for the 3000 foot runway 24 LOC DME, with very steep terrain immediately east. The closer I got to the airport, the more they tried to talk me out of landing there. My alternative was BGSF, which has the most consistent clear weather in the world. South winds 25 knots gust to 40, no fuel available, 500 bkn and one mile. I would not have minded using auto-gas, but the $8.28 per gallon fuel showed up somehow the next day. They are very careful with the fuel. You must sign a statement after you see their fuel tank sump sample with no water. The landing fees are $120. There is a ski lift next to the airport. Tourist Association fax 22710.

Texel (Texel International Airport – EHTX)

Vliegveldrestaurant Texel 

I recommend the Vliegveldrestaurant Texel on my, Texel International Airport, located on the island of Texel. I really recommend the famous "Tandemburger." Come and visit EHTX!

Strandhill, Sligo (Sligo Airport)

Ocean View Inn

All the free mussels and dillisk you can eat, on the beach just next to the runway. If you want something warm, head into town, about a ¾-mile walk up airport road, to the Ocean View Inn -- wonderful menu, reasonable prices, and good Guinness across the street at the Venue.

Pisa (Pisa Intl.)

Terminal Cafe

There is a restaurant inside the airport terminal. It is a normal airport-variety snack bar, Italian style, with sandwiches, pizza, pastries, drinks, etc.. The average cost of a meal is about 20,000 Lira. That is about $14.00. The kicker is that automobile fuel is running about $5.25 per gallon, and I assume that aviation fuel is proportionally higher. I only noticed about three to four general aviation aircraft during the entire month of March, and I think that those belonged to a charter outfit. I think that this is more of a $1,000 hamburger, or pizza, I should say.

Like most of Europe, public transportation is readily available at the airport. There are buses and a train station at the terminal. You can take a bus into town for about $3.00 round-trip and see some old building that is about to fall over. If it was in the U.S., it would be a parking lot by now. The Leaning Tower is worth the short trip. For about $10.00 round-trip, you can take a train into Florence.

A nice thing about Italy is that I tried many restaurants, both big and small, and did not find one that disappointed me. Also, there are a lot of beautiful buildings and scenery everywhere. I highly recommend a trip there.

Cancun (Cancun Airport)

Airport Restaurant

Of all the FBOs I've been to in Mexico, Cancun's is the best, five-star service and great cheeseburgers served from a little bar that seats six. Try to avoid going there between noon and 1:00 PM because all of the airport employees are eating there. Jet drivers get fueled first, but GA pilots are treated just as well.

Matamoros (Matamoros Airport)

Mariscos del Golfo 

If you go into Brownsville, park at Border A/C (has never charged parking, and I don't buy the $2.20 gas) or Hunt PanAm. They will sometimes drive you the eight miles to the bridge (give a $5.00 tip). Cabs wait at the bridge to bring you back for $8.00 if you bargain. Once you walk across the bridge, turn right and walk through a pedestrian opening in a tall white wall. This is the stop for the maxi-taxis and buses that charge 1 peso for adults and 1/3 peso for children. Bus transportation here is cheap, easy and reliable. Take the bus that says AZUL this route goes about 10 miles and stops at all the main attractions in town and ends at the Matamoros Airport entrance road. Get off at the Hotel Hernandez, like a Motel 6 for $25 per night or the Hotel Hardin $80.oo per night and very nice. Two blocks NW of the Hernandez is a seafood restaurant, Mariscos del Golfo, not fancy but clean. Two blocks west of the Hernandez is a local market not based on tourists. Four blocks east is the central bus station, I like to take the 40-minute bus ride to the beach. Six blocks south of the Hernandez on the way to Hotel Hardin is Los Potalles, a very fancy restaurant serving grilled meats. Further south on the AZUL route is a mall and a new McDonald's.

Mexicali, Baja (Mexicali Airport)

Airport Cafe 

Mexicali is a port of entry. When you have finished with officialdom and paid for gas, move to adjacent tie-down. Walk across the tarmac to the terminal just as if you owned the place. Keep trying doors until you find one unlocked; the farthest ones are your best bet. Go upstairs to a good place to have breakfast or lunch.

Monterrey (Monterrey Int'l Del Norte Airport - MMAN)

Enchilada Heaven 

The Mexican government prefers you stop at the nearest airport to where you cross the border. If flying from the Houston area, you can go direct to Monterrey because there is no airport with customs near where you cross the border. On this direct route, Houston to MMAN, you file a DVFR flight plan with "advise customs" in the remarks section. As you cross the border, call Monterrey Approach on 118.3. As you cross the 5,000-foot mountains about 30 north of the airport, they switch you to tower 118.6. When you report 10 miles out, they will generally clear you to land straight in on 20. After landing exit to the base of the tower and park inside one of the yellow circles for customs inspection. After an official looks at your plane, you go into the terminal just south of the tower and each person fills out a post card-sized declaration. They are available in English. Next, there is a yellow post with red and green traffic lights. They hold your bags next to the post and push a button. If green, you pass; if red, they ask to look in your bag. One in our group of five had his bag searched. I was told it is just a random counter. The pilot signs an entry and exit logbook (which is held by this very pretty young lady). Then each person fills out a tourist card. Each person should have one of the following documents: notarized copy of birth certificate, passport, or voter registration. One of my passengers only had a nonnotarized copy of a birth certificate, and it was OK.

Next, the pilot goes to the commandant's office just north of the tower with four documents: Mexican insurance, pilots license, airworthiness, and registration. The insurance can be obtained for as little as $80.00 per year at (800-423-2646). As you leave the next building, north is where you buy fuel and they will show you where to park. You can taxi or push the plane from the yellow circle across to the north corner of the ramp for parking. There are no ropes or places for ropes. I used my own chocks.

As you leave the terminal it is 50 feet to the main road to Monterrey, and a bus comes by every 15 minutes until 9:00 PM and charges about $1.00. You can call 800 numbers from most Mexican pay phones by dialing 95-800…

When you arrive at the airport for departure, you may need to go through the guard gate just south of the terminal if it is locked. You go to the same office and turn in your tourist cards. The pilot signs the same entry/exit logbook. Next to this office is the weather office where you can file a flight plan again. You advise customs in the remarks section. I chose to return via KLRD so I would not lose the westerly upper wind component. Also, that stupid balloon near KPSX is in my way if I go through KMFE.

The Mexican weather service wants you to be very specific about your departure time to help the U.S. track you. Next, take your flight plan to where you got your tourist cards and have it stamped. Now, one final trip to the commandant's office.

The Mexican officials were very helpful, friendly, and there were no fees of any kind (i.e., parking, customs, airspace, landing), none! It is well guarded. They are open seven days a week from 7:00

AM to 7:00 PM. They use the same time as CST year-round.

You can speed customs in LRD if you have a form 178 completed. They will charge you $25.00 for the annual processing fee sticker.

Mulege (Mulege Airport)

Hotel Serenidad 

Well, it's a bit out of the way, but Hotel Serenidad in Mulege, Mexico, has a GREAT pig roast every Saturday night, a tradition for many (30?) years. The flight down is very scenic. The hotel has a nice hard clay strip, I believe it is about 3,500 feet long. Mulege is 385 nm south of Mexicali, which is about 100 nm south of Palm Springs. Getting in and out of Mexico is easy. Weather is hard, VFR 98 percent of the year in the Baja.

AMENDMENTS

Mulege was great, but it was closed in late 1996, seized by a local Mexican ejito, or labor cooperative. It is tied up in court now, but it looks like it will never be reopened. Just a few miles north is Punta Chivato. It is a resort hotel with a hard-pack dirt strip and lots of Americano homes. The bar/restaurant is okay, and they are advertising a Saturday nightpig roast, since the Serenidad has closed. Give it a try.

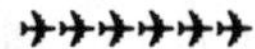

My hangar neighbor, Ray, just returned from Baja. Quite by accident he dropped in to Mulege to see what was going on and found that the Hotel Serenidad was just opening again under the original management. Ray said the rooms had been refurbished and the pool repaired and refilled. The grounds looked great, and the restaurant was up and running and tastin' good. No fuel (yet)!

Culebra (Culebra)

Happy Landings

There is a restaurant 20 feet outside the fence on Culebra, Puerto Rico. Not sure they even HAVE a menu. The best bet is to just say, "Bring me the best you have today."

The island is midway between St. Thomas, VI, and San Juan, P.R., about 40 nm from each and 15 miles due east of Fajardo.

Culebra runway is SE/NW straight into the trade wind for takeoff, about 3,000 feet. It is not a landing for the faint-of-heart. In 10 years, I've NEVER seen any aircraft depart or arrive to the NW, partly because of the almost constant tradewind and also because of the 450-foot mountain DIRECTLY at the western end of the runway. The islanders ease over the top of the mountain, line up with the runway, all but brush the treetops and immediately go full flaps, nose over, skim the trees and flare onto the runway. The alternative is to approach from the large bay on the northwest end of the island, follow the road (only one) over the saddle, STAY LOW to the end of the runway, turn southeast and flare.

DON'T EVEN THINK ABOUT NIGHTTIME OPERATION!

It is an absolutely delightful, nontouristy island. Don't attempt to rent a car. There are only three or four there, and the operator doesn't have a phone. You are just supposed to tell someone to tell him you are there. Besides that, there are only two roads on the island. The only intersection is at the airport, and the first car/truck that goes by will pick you up and drop you wherever you want to go. It will be assumed to be free. If you have a gang of luggage or if it takes a couple of trips, tell them you'll buy a beer next time you see them. You will, and it would be a nice gesture. Do NOT attempt to pay.

My 76-year-old mom rented a car the first two weeks we were there, $200 a week, and managed to log 7.5 miles. When we cleared the island on the ferry the last time, it took three vehicles. All of them stopped, loaded junk aboard, tooled to the ferry and off-loaded it and us and waved bye. El Batey restaurant is nicer, about 1/4 mile toward "town." It is hard to spend $10.00 on a dinner.

Snorkelers will find the reefs just like the VI next door. However, the most popular beach, two miles of gorgeous white sand with sea grape and palm trees, will have two dozen people on it on the busiest day of the year. Normally there is no one, or perhaps two or three. We left diving gear, cameras, and radios in a pile on a blanket, went to Dewey for beer, came back seven hours later, and found someone had pulled our stuff up from the tideline and left a "howdy" in the sand. Crime is nonexistent. English about 40 percent of the 3,000 population: about 500 "yachties," 500 displaced New Englanders and 2,000 or so "old style" Puerto Ricans. There are about 20 rooms for rent sprinkled around.

AMENDMENT

The restaurant adjacent to the airport is called Happy Landings, a name that seemed ironic when it came to light in an Associated Press wirephoto seen worldwide. Hurricane Marilyn had deposited a private plane upside down on HL's roof.

Most of the HL's menu is deep-fried Puerto Rican-style cooking--empanadas, meat pies,etc.

United Kingdom

Barton Aerodrome, Manchester

Starvin Marvin's

Starvin Marvin's, Salford Quays is an authentic 60's American roadside diner with great food and reasonable prices. Supposedly, it was fetched in two halves from its original site by a road in Alabama and reassembled in slightly-less-sunny Manchester. Typical American fare. An easy walk from the field.

Shobdon (EGBS - EGBS)

Shobdon Airport Cafe 

Shobdon, EGBS, is 6 nm west of Leominster, on the Welsh borders. I know it's a long way from the American Southwest, but for those flying in the UK, you won't find a friendlier airfield. The restaurant is warm, satisfying and inexpensive, and the outdoor seating has a clear view of the runway, essential in my opinion!

Your Name ______________________ Date ______________

Restaurant Name ______________________________

City ______________ State ________ Country ________

Airport Name ______________________ Identifier ______

Your Report

____

____

____

____

____

____

____

Hours of Operation ______________________________

Restaurant Phone # ______________________________

FUEL Price ________

100LL ______ Jet A ______ Mogas ______ 80 Octane ______

Fuel Comments

____

____

____

____

____

The Rating Points

Location
Ramp Side = +2
Short Walk = +1
Crew Car = 0
Cab/Bus = -1
Long Walk = -2

Food
Perfect = +2
Average = +1
Unsat. = -1
Awful = -1

Price
Top Value = +1
Average = 0
Be Warned = -1

Service
Perfect = +1
Average = +0
Rude = -2

Ambiance
Special = +1
Average = 0
Bad Vibes = -1

Your Rating ________

Lake Texoma Resort Park
Kingston, OK
580 / 564-2311

Exchange this coupon at the front desk for another good for a 15% discount off pilot's food purchases. FREE transportation provided!

Offer expires June 30, 1999

Grand Casino Avoyelles
711 Grand Blvd. Marksville, LA 71351
800 / 784-7760

Present this coupon at the *Marketplace Buffet* to receive TWO FREE BUFFETS! (All-you-can-eat; Over 95 items)

Must be 21 years of age. Offer expires December 30, 1999

Silver Wings Restaurant & Bar
500 James Fowler Rd. Santa Barbara, CA 93117
805 / 964-7793

Upstairs Santa Barbara Airport Terminal, this coupon entitles pilot and passengers "20% off anything at our restaurant."

Offer expires December 30, 1999

Howard's Pub & Raw Bar Restaurant
½ mi. So. of the airfield on NC Hwy 12
Ocracoke Island, NC
919 / 928-4441

Receive ½ off pilot's meal with this coupon!
FREE transportation provided: Call from the National Park Service gazebo at the airstrip.

No expiration date.

The HEARTLINE CAFÉ
1610 W. Hwy 89A Sedona, AZ 86336
520 / 282-0785

10% discount to pilot presenting this coupon!

Offer expires December 31, 2000

Castle Air Museum
Flight of Fancy Grill
Atwater, CA 95301
209 / 723-2178

Buy one B-24 Liberator Hamburger Combo and redeem this coupon to Get the second at half price!

Offer expires December 31, 2000